THE FUTURE FOR RENEWABLE ENERGY2

PROSPECTS AND DIRECTIONS

EUREC Agency

JAMES
X
JAMES

Published by James & James (Science Publishers) Ltd
35–37 William Road, London, NW1 3ER, UK

A catalogue record for this book is available from the British Library.

ISBN 1 902916 31 X

Printed in the UK by The Cromwell Press

Cover photos: Six 9 kWe dish/Stirling units at the Plataforma Solar de Almeria (CIEMAT-PSA,
Spain); solar home system, Lesotho (IT Power); an experimental sweet sorghum plot in Sicily
(Dr V Sardo); wind farm near Zaragoza, Spain (MADE); retrofitted students' hostel in Dornbirn,
Austria (architect Heim Müller); computer generated image of an array of axial flow tidal current
turbines (Marine Current Turbines Ltd).

CONTENTS

Authors and acknowledgements   **xi**

Preface   **xiii**

Introduction   **xv**

1 Biomass   **1**

1.1   A magnificent gift   1
1.2   Introduction   2
  1.2.1   Potential and strategic relevance   2
  1.2.2   Conceptual clarification   3
  1.2.3   Structure of the biomass chain   3
  1.2.4   Biomass potential   5
  1.2.5   Policy environment   6
1.3   Present situation   6
  1.3.1   Present bioenergy contribution   7
  1.3.2   Techno-economic aspects   8
  1.3.3   Techno-political aspects – research policy   8
  1.3.4   Socio-technical aspects   10
1.4   Goals for research, development and demonstration   11
  1.4.1   Feedstock-related research   12
  1.4.2   Conversion-related research   12
  1.4.3   Research on bioenergy end-uses   12
  1.4.4   Research on bioenergy systems   12
1.5   A road to a stronger market   14
Annex – Biomass conversion technologies   14

2 Future research and development in photovoltaics   **25**

2.1   Strategic summary   25
2.2   Introduction: Photovoltaics in a nutshell   26
  2.2.1   Principle of operation   26
  2.2.2   Production technology   27
  2.2.3   Wafers and junction formation   28

2.2.4   Cells   28
2.2.5   Modules   29
2.2.6   Photovoltaic systems components   29
2.2.7   Applications   30
2.2.8   Potential and features   30
2.2.9   Current cost and cost reduction potential   31
2.3   Achievements   32
2.3.1   Current situation   32
2.3.2   Updated goals   34
2.3.3   Main barriers to overcome   34
2.4   Roads to a stronger market   37
2.4.1   Setting the legal, fiscal and political framework   37
2.4.2   Advertising the benefits   38
2.4.3   Building the 21st century image   38
2.5   Goals for research, development and demonstration (RD&D)   38
2.5.1   The way forward: the R&D roadmap   38
2.5.2   Crystalline silicon solar cells   39
2.5.3   Thin-film crystalline silicon   45
2.5.4   Amorphous silicon   47
2.5.5   Thin-film polycrystalline technologies   49
2.5.6   III-V cells   52
2.6   Advanced research and cross-fertilization with other
        technology development   54
2.6.1   New research approaches for highly efficient photovoltaic conversion 55
2.6.2   Organic and polymer cells   55
2.6.3   Modules   56
2.6.4   Concentrators   57
2.7   Systems integration and technologies   58
2.7.1   Technology status of components   58
2.7.2   Generic R&D needs for systems   59
2.7.3   Goals for technology development: systems integration   59
2.7.4   Goals for technology development: electricity storage   62
2.7.5   Goals for technology development: components   62
Annex   63

3.   **Small hydro power**   **65**

3.1   Introduction   65
3.2   Potential and strategic summary   66
3.2.1   Basic principles   68
3.2.2   Experiences to date   69
3.2.3   Present state-of-the-art   70
3.2.4   Industry and employment   71
3.2.5   Distribution of small hydro power within Europe   72
3.3   Goals for research, development and demonstration   72

3.3.1   Long-term goal (by 2005–2010)   72
3.3.2   Medium-term goal (by 2000–2005)   72
3.3.3   Short-term goals (immediate)   72
3.3.4   R&D roadmap   73
3.3.5   Data collection on resource and on existing installations   74
3.3.6   Technical RD&D   74
3.3.7   Development of standard specifications for design and installation packages
        and harmonization of engineering standards   76
3.3.8   Evaluation of the real environmental effects   76
3.4   Roads to a stronger market   76

4.   Energy efficient solar buildings   79

4.1   Introduction   79
4.2   Present situation   84
4.2.1   Residential buildings   85
4.2.2   Non-residential buildings   92
4.2.3   Retrofits in the existing building stock   102
4.3   Goals for R&D   104
4.3.1   Components   105
4.3.2   Planning tools   109
4.3.3   Innovative educational colloquia   109
4.3.4   Monitoring and evaluation   110
4.4   Roads to a stronger market   110
4.4.1   Benchmarking: a call for new building codes   112
4.4.2   Quality control   112
4.4.3   Further research   113
4.4.4   Retrofitting: the road not taken   113

5.   Solar thermal power plants   115

5.1   Introduction, potential and strategic summary   115
5.2   Achievements, present situation, main barriers and vision for the way
      forward   119
5.2.1   Parabolic trough systems   119
5.2.2   Central receiver systems   123
5.2.3   Parabolic dish systems   128
5.3   Goals for research, development and demonstration (RD&D)   132
5.3.1   Commercial-scale projects   132
5.3.2   Cost reduction   132
5.4   Roads to the market   133
5.4.1   Parabolic trough systems   134
5.4.2   Central receiver systems   135
5.4.3   Parabolic dish systems   136

6.   Wind energy   138

6.1   Introduction, potential and basic strategy   138
  6.1.1   Introduction   138
  6.1.2   Potential: the European wind energy resources   139
  6.1.3   Basic strategy   141
6.2   Present situation and achievements   144
  6.2.1   Status of the technology   144
  6.2.2   Markets   150
  6.2.3   Industry and employment   152
  6.2.4   Economy   152
  6.2.5   Non-technical implementation issues   156
6.3   Goals for R&D   159
  6.3.1   Introduction   159
  6.3.2   Cost reduction of wind energy   159
  6.3.3   Increasing the value of wind energy   164
  6.3.4   Finding new sites   164
  6.3.5   System development   165
  6.3.6   Reduction of uncertainties   167
  6.3.7   Reduction of environmental effects   168
6.4   Roads to market   169
  6.4.1   Land-based wind energy systems in industrialized countries   169
  6.4.2   Land-based systems in developing economies   170
  6.4.3   Wind energy systems for remote areas   170
  6.4.4   Offshore applications   170
  6.4.5   General   171

7.   Integration of renewable energy sources into energy systems   173

7.1   Introduction and strategic summary   173
7.2   Strategic considerations   174
  7.2.1   Policy and politics   174
  7.2.2   Priorities and schedules   175
  7.2.3   Key players   176
7.3   Present situation   176
  7.3.1   RE technologies in Europe   176
  7.3.2   Integrated RE systems   177
7.4   Integration issues   178
  7.4.1   Technical issues   178
  7.4.2   Non-technical issues   182
7.5   Goals for a European RE integration strategy   187
  7.5.1   The vision   187
  7.5.2   Strategic goals   188
  7.5.3   RD&D areas and tasks   188
  7.5.4   Road to the market: establish the appropriate legislative framework   192

8.   Renewable energy technologies for developing countries   194

8.1   Introduction   194
8.2   Present situation   195
   8.2.1   Energy demand   195
   8.2.2   Renewable energy technologies   197
   8.2.3   Rural electrification   202
8.3   Goals for RD&D   204
   8.3.1   Biomass   204
   8.3.2   PV   204
   8.3.3   Small hydro   205
   8.3.4   Solar thermal power   205
   8.3.5   Wind   205
8.4   Road to market   205

9.   Ocean energy   207

9.1   Introduction   208
9.2   The potential of marine energy: strategic summaries   208
   9.2.1   Tidal/marine currents   208
   9.2.2   Wave energy   209
   9.2.3   OTEC (ocean thermal energy conversion)   210
   9.2.4   Tidal barrages   210
   9.2.5   Salinity gradient/osmotic energy   211
   9.2.6   Marine biomass fuels   211
9.3   Present situation   211
   9.3.1   Tidal/marine currents   211
   9.3.2   Wave energy   213
   9.3.3   OTEC (ocean thermal energy conversion)   215
   9.3.4   Tidal barrages   215
   9.3.5   Salinity gradient/osmotic energy   216
   9.3.6   Marine biomass fuels   216
9.4   Economic aspects   217
9.5   Environmental aspects   217
9.6   Goals for research, development and demonstration   218
   9.6.1   Generic goals for primary offshore technologies   218
   9.6.2   Realistic goals for tidal/marine current energy development 218
   9.6.3   Realistic goals for wave energy development   219
9.7   Roads to a stronger market   220

10.   Solar process heat   222

10.1   Introduction, potential and strategic summary   222
10.2   Achievements, present situation, main barriers and vision for the way
        forward   224

10.2.1 Main barriers   226
10.2.2 Vision for the way forward   226
10.3 Goals for research, development and demonstration (RD&D)   226
    10.3.1 Successful demonstration of current technology   226
    10.3.2 Development of numerical design and assessment tools for solar process heat
        system integration   227
    10.3.3 Development of improved components and subsystems   228
10.4 Roads to the market   228

11.   Solar chemistry   231

11.1 Introduction, potential and strategic summary   231
11.2 Achievements, present situation, main barriers and vision for the way
    forward   232
11.3 Goals for research, development and demonstration (RD&D)   236
11.4 Roads to the market   237

Appendix A The EUREC Agency   240

The European Renewable Energy Centres Agency   240
EUREC Members   242

Appendix B   246

Physical units and conversion factors   246
Prefixes   246

Index   247

AUTHORS AND ACKNOWLEDGEMENTS

Chapter 1: Biomass

Author: Emmanuel G. Koukios (NTUA, Greece)

Acknowledgements: Hubert. Veringa (ECN, The Netherlands), Luisa Delgado Medina (CIEMAT, Spain), Herman den Uil (ECN, The Netherlands), Jacob Bugge (Folkecenter, Denmark), Nicola Pearsall (NPAC (University of Northumbria), UK)

Chapter 2: Future research and development in photovoltaics

Author: Heinz Ossenbrink (JRC, Italy)

Acknowledgements: Johan Nijs (IMEC, Belgium), Wim Sinke, Paul Wyers (ECN, The Netherlands), Nicola Pearsall (NPAC (University of Northumbria), UK), M. Lux-Steiner (HMI Germany), E Dunlop (JRC Italy), H Gabler, F Oster (ZSW, Germany), Thierry Langlois d'Estaintot (European Commission DG Research), Bill Gillett (European Commission DG TREN)

Chapter 3: Small hydro power

Author: Peter Fraenkel (IT Power Ltd, UK)

Acknowledgements: Jochen Bard (ISET, Germany), Philippe Schild (European Commission DG Research), Tom Thorpe (AEA, UK)

Chapter 4: Energy efficient solar buildings

Author: C F Reinhart (Fraunhofer Institute for Solar Energy Systems, Germany)

Acknowledgements: J Luther, K Voss, V Wittwer (Fraunhofer Institute for Solar Energy Systems, Germany), Th Van der Weiden (ECOFYS, The Netherlands), N Fish (STZ-EGS, Germany), M Santamouris (University of Athens, Greece), W Weiß (AEE Intec, Austria)

Chapter 5: Solar thermal power plants

Authors: Manfred Becker, W Meinecke, M Geyer, F Trieb (DLR, Köln, Stuttgart and Almería, Spain), Manuel Blanco, M Romero (CIEMAT, Madrid and Almería, Spain), A Ferrière (CNRS, France)

Acknowledgements: Philippe Schild (European Commission DG Research), Arthouros Zervos (NTUA, Greece)

Chapter 6: Wind energy

Authors: Jos Beurskens (ECN, The Netherlands), Peter Hjuler Jensen
                        (Risø, Denmark)
Acknowledgements: Peter Hauge Madsen, Erik Lundtang Pedersen (Risø, Denmark),
                        Jens Peter Molly, H. Seifert (DEWI – Germany), Andrew Garrad
                        (Garrad Hassan Partners), Gijs van Kuik (TU Delft, The
                        Netherlands), Jürgen Greif (European Commission DG
                        Research)

Chapter 7: Integration of renewable energy sources into energy systems

Authors: Arthouros Zervos, Danae Diakoulaki (NTUA, Greece),
                        Didier Mayer (ARMINES, France)
Acknowledgements: A Louche (E²H, Corsica, France), Manuel Sanchez Jiminez
                        (European Commission DG Research)

Chapter 8: Renewable energy technologies for developing countries

Authors: Bernard McNelis (IT Power, United Kingdom), G. van Roekel
                        (ECN, The Netherlands), Klaus Preiser (FhG-ISE, Germany)
Acknowledgements: Wolfgang Palz (European Commission, DG SRC), P. Lasschuit (ECN,
                        The Netherlands), K. Weiner (FhG-ISE, Germany)

Chapter 9: Ocean energy

Author: Peter Fraenkel (IT Power, UK)
Acknowledgements: Tom Thorpe (AEA, United Kingdom)

Chapter 10: Solar process heat

Authors: K Hennecke, B Hoffschmidt, W Meinecke (DLR, Köln, Germany),
                        M Blanco (CIEMAT, Spain)
Acknowledgements: Philippe Schild (European Commission DG Research), A Zervos
                        (NTUA, Greece)

Chapter 11: Solar chemistry

Authors: Karl Heinz Funken, C Sattler, C Richter, R Tamme (DLR, Köln
                        and Stuttgart, Germany), J Blanco (CIEMAT, Spain), Jacques
                        Lédé (CNRS-LSGC, France)
Acknowledgements: Philippe Schild (European Commission DG Research),
                        Emmanuel G Koukios (NTUA, Greece)

PREFACE

In Europe, and industrial countries in general, environmental concerns are the driving force behind the development of renewable energy sources. It is generally accepted that the continuation of existing energy structures is not sustainable, enhancing phenomena such as global warming and causing climate change. Renewable energy technologies can make significant contributions to reducing greenhouse gas emissions from the use of fossil fuels. The key issues for increasing the use of renewable energy actually relate to making these technologies cost-competitive and integrating them into the existing energy system.

However, for more than two billion people living in villages and rural areas in the world's developing countries energy is a matter of pure survival. They have no access to electricity. In many places women and children spend many hours each day collecting wood or dung as fuel for basic energy needs. Low energy conversion efficiencies combined with fast growing energy needs are the basis of ecological problems such as deforestation and erosion. Meeting the energy demands of these growing populations in developing countries with fossil-fuel-based systems would increase ecological damage on a global scale.

In Europe, the White Book for renewable energy sources has set the first scenarios for renewables to contribute to the total energy supply. It is important that the target of a contribution of 12% of the total primary energy consumption in Europe is met by the year 2010 and that society adapts to integrate renewables. In some EU countries and for some renewable energy technologies the progress is far beyond the projected targets. For others, development is yet to start. Legal and institutional frameworks are under development. They are mainly based on environmental concerns, but factors such as security of supply and socio-economic development also play an important role. A concrete example is the recently adopted European Directive for electricity from renewables in the free European market.

In recent years the renewable energy sector has seen double-digit growth figures. However, in absolute figures, the renewables business sector is still small. It is a high-tech sector with emerging technologies depending heavily on the results of research and development. Technological achievements of today are the result of the vision, persistence and hard work of many researchers and engineers over more than 20 years. The future success of renewable energy technologies can only be assured if efforts in research, development and demonstration are maintained or strengthened for at least the same period. EUREC brings together most of the European research institutes renowned for their expertise in the field of renewable energy sources. This book represents the ideas and visions of over 1000 researchers throughout Europe. We hope that *The Future for Renewable Energy 2* will constitute a solid basis for any national or European research policy in the renewable energy field.

TRIBUTE TO PROFESSOR ROGER VAN OVERSTRAETEN

Professor Roger Van Overstraeten had a passion for solar energy and renewable energy sources. He started intense R&D activity on solar photovoltaic cells in the early 70s. The main aim of his work was to improve efficiency and at the same time reduce the cost of solar cell panels so that energy could be provided to people living in remote areas at a reasonable cost. At that time he was already assisting the European Commission as advisor of its R&D activities in the field of photovoltaics, being part of one of the first technology R&D programmes besides Euratom, long before the Framework Programmes were set up.

Professor Van Overstraeten was one of the architects of the annual European Photovoltaic Solar Energy Conferences. He was the first winner of the prestigious Becquerel Prize. This prize was set up by the European Commission on the occasion of the 150th anniversary of Becquerel's discovery of the photoeffect which was the basis for photovoltaics. The prize was awarded to Professor Van Overstraeten in 1989, at the 9th EC Photovoltaic Energy Conference in Freiburg. Since then, the prize has been given to outstanding personalities in photovoltaics at each of the European PV conferences.

Professor Van Overstraeten was the driving force behind the creation of the Inter-university Microelectronics Centre (IMEC) in Leuven, Belgium. Even in the early days, the work he supervised at K.U.Leuven and at IMEC was outstanding on a European and international level. Today the PV group at IMEC has become one of the most prestigious European teams for crystalline silicon solar cells and plays a crucial role in several pioneering high-technology PV projects.

Professor Van Overstraeten represented IMEC as founding Member of the EUREC Agency E.E.I.G. He was EUREC president from 1993 until 1996. Prof. Van Overstraeten died on 29 April 1999.

INTRODUCTION

EUREC stands for European Renewable Energy Centers. The EUREC Agency was set up in 1991 as a European Economic Interest Grouping (EEIG) to provide a forum for interdisciplinary co-operation between renewable energy research organizations. It now has about 50 members forming a strongly collaborating group of independent organizations, joining forces with the common goal of fostering the uptake and integration of renewable energy in society. Appendix A provides further details on EUREC and its objectives as an organization. The latest information on the EUREC member organizations and their activities can be found on the website http://www.eurec.org.

Energy is still an important issue globally. EUREC and its members are totally devoted to research, development and demonstration (RD&D) in this area. Besides performing research, developing and demonstrating in practice renewable energy components and systems, EUREC member institutes are analysing the market for renewable energy technologies and setting up roadmaps in order to organise their RD&D activities in the best way. This book reports on the findings of these activities. EUREC, due to its European-wide representation and coverage of all renewable energy technologies, is a good source for this evaluation.

The first edition of EUREC's *The Future for Renewable Energy* reflected the situation of renewable energy sources and the technology for their exploitation in 1995. This second edition reflects the situation in the year 2000. The updating of the book has to be considered as an ongoing process; the chapters can be considered as a 'snapshot' at a certain point in this process.

In this edition we do not discuss the technical potential of each individual technology in detail as was done in the 1995 edition. The potential for supplying energy to meet – if necessary – 100% of the global demand has been proven. Quantitative evaluation in terms of resource calculation or figures on potential and installed capacity is available in multiple other recent publications.

The chapters of this book cover the different renewable energy technologies, giving a brief description, recent achievements, goals for a strong and coherent RD&D policy and a vision for the way forward. There are also separate chapters on the issues of integration and developing countries, which are both of major importance for renewable energy in general. The chapters in the book thus cover:

- biomass
- photovoltaics
- small hydro
- solar buildings
- solar thermal power stations
- wind energy
- integration
- developing countries
- other renewables including ocean energy, solar process heat and solar chemistry

As EUREC has chosen to concentrate on its core competencies, not all technologies will be discussed in this book; for example, geothermal energy is not covered. Also, the book does not elaborate on issues beyond energy in other areas of sustainability, such as material consumption or environmental efficiency.

In compiling the chapters for this book, EUREC's main objectives were as follows:

- **To provide an evaluation of the current achievements of renewable energy.** This edition contains an evaluation of what has been achieved within the field of renewable energy in the course of the past few years.
- **To set goals for research, development and demonstration (RD&D) – research and development goals for the future follow from the analysis of the present situation.** General goals are cost reduction, higher conversion efficiencies, new products, integration and aesthetics. It is generally recognized that a very strong 'innovation trend' exists. In this trend, research is performed only as a function of industrial innovation and once a technology is on the market, it is thought that no further research is necessary. The danger of this trend becoming too strong is that application-oriented fundamental research will be less important or totally excluded. The objective of this book is to emphasize that industrial innovation is 'consuming' research results, so that the latter have to be produced continuously. The book presents renewable energy research visions, in order to demonstrate the necessity of research and development apart from innovation.
- **To show roads to a stronger market.** For each technology sector the book examines what should be done to foster technology development and market introduction in renewable energies.

The main statements in this book have been discussed with the major renewable energy industry associations such as EWEA – the European Wind Energy Association, ESHA – the European Small Hydro Association, ESIF – the European Solar Industry Association, EPIA – the European Photovoltaic Industry Association, and EUBIA –

the European Biomass Industry Association. Since summer 2000, EUREC has been sharing offices in Brussels with some of these associations, thus joining forces in support of renewables.

The ongoing process of de-regularization of the European energy market forces the energy sector to rapidly acquire expertise, with competition in electricity generation, distribution and trade. This development contains risks as well as opportunities for the deployment of renewables. Competition puts pressure on energy price levels, making it more difficult for renewables to compete with traditional energy sources. However, new players arrive in the competitive market, further supporting the trend towards more decentralised power and favouring an increased integration of renewable energy sources. This development will lead to many technological as well as regulatory challenges and will require a further development of legal and policy framework conditions. In any case, a really liberalised market should be based on a level playing field, taking into account the 'external costs' of different energy sources (including CO_2 effects), as well as socio-economic factors such as the potential for security of energy supply, export possibilities and employment.

As long as external costs are not taken into account, governmental strategies are needed for a limited time for market introduction of renewable energy technologies. The time horizons necessary differ depending on the type of the renewable energy source. Additionally, energy strategies and technologies employed show strong regional differences. Examples are climatisation of buildings: heating in the north, heating and cooling in the Mediterranean area, dehumidification in tropical climates (export of technologies from Europe). Thus there is no 'single renewable energy strategy' for the whole of Europe. However, in some areas a 'single strategy' or regulation for the whole of Europe is appropriate or even necessary: examples could be the 'internalisation' of the external costs in energy prices and the introduction of 'codes' – especially building codes – adapted to a European-wide 'sustainable energy service strategy'.

The capital intensity of renewable energy technology constitutes a big hurdle with respect to introduction into today's market. The high investment costs are compensated by non-fuel-consumption during the time of operation (with the exception of biomass). The fact that the costs of renewable energy supply are mainly due to investments at the very beginning of the installations is on the other hand an advantage: there is a clear basis for the calculation of future energy cost. Financial risks are thus low.

Finally it has to be stressed that targeted education and training schemes for scientists, engineers, craftsmen, architects, contractors, financial agents, etc. are essential in order to ensure the availability of sufficient 'human resources' for the future deployment of renewable energy sources.

Prof. J. Luther   K. Derveaux
President of EUREC   Secretary General

Brussels, Belgium, November 2001

1. BIOMASS

E G Koukios (NTUA)[1]

1.1 A MAGNIFICENT GIFT

An alien spaceship lands somewhere in Europe; its crew members disembark and their leader makes the following speech to the Europeans crowded to welcome the extraterrestrial beings on Earth:

> People of the Earth, we are coming as friends and, to prove our friendship, we are bringing you a magnificent gift. Based on our long observations from space, we have noted that your part of the Earth – what you call 'Europe' – is plagued by a number of serious problems. Our gift is intended to help you face most of these problems at once, by making it possible for you to 'trap' large amounts of solar energy and store it in usable forms for as long as you plan and wish. This 'trapping' is accompanied by the consumption of carbon dioxide and the production of extra oxygen. If and when you decide to use it, a great number of possibilities will be open to you, including energy (heat and electricity) and fuels, fibres and other materials, chemicals, as well as various useful fine and bulk products. All these products and uses are environmentally friendly and will hopefully be able to replace some of your ecologically troublesome activities. Local people will be able to find stable employment in the new production and conversion units, your agriculture could be rationalized, and your rural regions could enter a period of new, balanced growth. Each European area will be able to choose the form of the gift that best suits its needs and strategies, and its peaceful complementarity with trade partners and neighbours.

The gift of our 'extraterrestrial' visitors is, of course, biomass.[2] This chapter formulates a hypothetical response of the Europeans to that 'magnificent gift', taking into account its opportunities and obstacles, and avoiding over-simplifications like 'it's too expensive' or 'it's not yet technically mature', which still dominate European debate. As the recent Biomass Conference of the Americas[3] concluded, the main role of an updated biomass strategy should be to make possible the 'harvesting' of the potential benefits of non-food bioresource uses. In the USA, authorities and other major players are facing this challenge; Europeans should also be mobilized to receive a 'gift' which, if it had not existed, we would have prayed for.

Figure 1.1 Transportation and handling of agricultural residues of bioenergy production could build upon the significant relevant experiences with food crops. Photo in a Greek rural community. (Prof. E G Koukios)

1.2   INTRODUCTION

1.2.1 Potential and strategic relevance

Although all renewable energy sources possess unique characteristics as part of their 'personality', biomass definitely leads the way in peculiarities, examples of which include:

*   It is notoriously complex.
*   There are mixed messages, e.g. high promises vs poor feasibility figures of large-scale applications.
*   There are disappointingly low market penetration rates, despite some local and regional success stories in Europe.

To generate an overall view of the complex biomass[4] territory is not easy. Studies of this type need to combine various types of expertise and experience, working across well-established boundaries, at the risk of not only simple 'mistakes', but even serious violations of scientific/technical disciplines outside the author(s)' expertise.[5]

Consequently, in order to arrive at reliable and truly useful results, this chapter is structured along methodological lines, including

*   an upgraded role for the other EUREC members as advisers and/or reviewers[6]
*   a comprehensive mapping of the complex biomass/ bioenergy territory, involving novel 'tools'
*   a market-oriented presentation style, focusing on the energy vectors generated from biomass, rather than the feedstocks and/or conversion technologies.[7]

Figure 1.2 Energy crops, such as sorghum, could be competitively grown by European farmers to replace conventional crops that are in excess. Photo of an experimental sweet sorghum plot in Sicily. (Dr V Sardo)

Due to their importance for the biomass research community, a strategic overview of the major conversion technologies is included as an Annex.

In the remaining parts of this introductory section we present the what, the why and the how of an EU biomass strategy:

- definition of the subject area, including all terms and concepts, to enable their use with greater precision
- definition of the three major 'compartments' of the biomass chains: resources, conversion technologies and end products, emphasizing the dominant role of the latter for strategic analysis
- assessment of the EU's biomass potential
- analysis of the strategic importance of biomass, as seen within several policy and socio-economic frameworks.

1.2.2 Conceptual clarification

In Table 1.1 we present a summary of the main terms and concepts used in the biomass field, with a few remarks on their exact meaning and value. To help the reader, the terms recommended for use are printed in bold.

1.2.3 Structure of the biomass chain

A cornerstone of the biomass conceptual territory is the distinction between the three main steps of the biomass chain, i.e.

resources → conversion → end-uses

Table 1.1 Clarifying the biomass language

Term	Meaning	Evaluation
Biomass	Biological mass, living or recently dead, in cellular form. Used also for some products, e.g. fuelwood	Simple term for the whole non-food spectrum. No clear indication of products
Phytomass	Plant biomass	Clear term, though used in a limited part of the field
Bioresource	Biological resources	Clear for biomass sources
Biomass energy	Energy from biomass	Unclear term, to be avoided. Confusion with other terms
Bioenergy	Energy from biomass	Clear, recommended term
Bioheat	Thermal energy from biomass	Clear, recommended term
Bioelectricity	Electrical energy from biomass	Clear, recommended term
Biofuels	All fuels generated from biomass, including solid (fuelwood, pellets), liquid (bioethanol, biodiesel, bio-oils), and gaseous (biogas, other gases)	Not to be limited to a class of these fuels, usually liquid ones for transportation. Recommended only if well specified (as on the left)
Biogas	A mixture of methane and CO_2 generated by anaerobic digestion of biomass (a bio-process)	Not very clear term, as it could cover all biomass-derived gaseous fuels
Bioethanol	Ethanol biologically produced from biomass-derived feedstocks	Term to distinguish from the petrochemically derived, chemically same, ethanol
Bio-diesel	Liquid fuel derived from vegetable oils, as diesel fuel substitute	Possible confusion with bio-oils if used in the same way
Bio-oils	Liquid output of biomass pyrolysis, of possible energy/chemical value	Not clear term – could cover all biomass-derived liquids
Bio-coal	Charcoal or product of biomass carbonization	Unclear – not to be used
Wastes	In connection to biomass, organic materials generated from various sources	Not clear, negative value term covering both biomass-related/unrelated outputs
Residues	Usually used for residual biomass left in the field or the agro-industry	Used by FAO, not negative but not clearly distinguished from process wastes
Byproducts, co-products	Could be used to characterize all types of biomass-related, waste and secondary flows of materials	Positive terms; if well specified, recommended for use
Biochemicals	Chemicals from biomass or produced by bio-processes	Unclear but useful term – should be specified
Biomaterials, Bioproducts	Materials and other products from biomass or by processes	Same situation as with bio-chemicals

The emphasis should be on end-uses, which, in a demand-driven economy, dictate the structure of the chain, i.e. which resources to use and which conversion processes to employ in order to produce the specific products or services in demand.

In addition to the three steps, the two main interfaces can be defined, i.e.

- resource/conversion interface, dominated by feedstock logistics and suitability
- conversion/end-use interface, where customer requests could be accommodated.

Several of the major barriers for the market deployment of bioenergy are connected with issues resulting from these two interfaces, particularly the former.

1.2.4   Biomass potential

The main biomass resources can be listed as follows:

- short rotation forestry (willow, poplar, eucalyptus)
- herbaceous ligno-cellulosic crops (miscanthus)
- sugar crops (sugar beet, sweet sorghum, Jerusalem artichoke)
- starch crops (maize, wheat)
- oil crops (rape seed, sunflower)
- wood wastes (forest residues, wood processing waste, construction residues)
- agricultural residues and wastes (straw, animal manure, etc.)
- organic fraction of municipal solid waste and refuse
- sewage sludge
- industrial residues (e.g. from the food and the paper industries).

Current and future available biomass resources in the European Union are given in Table 1.2.

Table 1.2 Biomass potential in the EU

Raw material	Current resources Mt (dry)/year	Future resources Mt (dry)/year
Co-products of other activities:		
Wood wastes	50	70
Agricultural residues	100	100
Municipal solid wastes	60	75
Industrial wastes	90	100
Dedicated land for biomass:		
Short rotation forestry	5	75–150
Energy crops	–	250–750
Total biomass	200	1000
(Total bioenergy, Mtoe)	(80)	(400)
(% current EU primary energy)	(5–6%)	(25–30%)

It can be seen from Table 1.2 that, in the long term, energy crops grown on land set-aside from agriculture could be a very important biomass fuel source. However, co-products of other production activities are the major biomass sources and are the priority feedstocks for energy production. There is also an added environmental benefit in using currently available secondary flows, such as municipal solid waste and sewage sludge as raw materials, as these are potential pollutants.

1.2.5 Policy environment

Another aspect of the complexity of biomass issues is their strong interrelations with multiple policy frameworks, in which biomass should be placed and examined. Table 1.3 summarizes the main policies that affect biomass questions and can be affected by them, and their main types of effect.

Table 1.3 Policies relevant to biomass*

Policies	Economic effects	Technology effects	Environmental effects	Societal effects
Energy	I	2	2	3
Industrial	I	2	3	2
Agricultural	I	3	2	I
Environmental (including greenhouse gases)	2	2	I	2
Research	2	I	2	3
Transport	3	2	2	3
Social (employment)	2	3	3	I
Regional/Rural	2	3	2	I
Education	3	2	2	I
Consumer	I	2	2	I
World development	3	2	2	3
International trade	3	2	2	2

* I = high relevance, 2= medium relevance, 3= low relevance

- Biomass-related strategic decisions have to be taken in more than a dozen different policy areas, as they are both affected by and may affect them.
- If not appropriately co-ordinated, this matrix of interactions could hinder potential applications, particularly novel ones.
- It appears possible to employ biomass utilization as a 'lever' to achieve multiple desirable effects in several policy areas.[8]
- Biomass strategies that fail to recognize the situation illustrated by Table 1.3 are not able to solve particular problems as they do not permit a comprehensive view of the whole system, thus limiting the chance of any viable solutions.

1.3 PRESENT SITUATION

The present use of bioresources for energy in the EU is described according to the major, market-based classification scheme of 'bioheat, bioelectricity, biofuels' introduced in the previous section.

Built around this axis, the 'mapping' of the biomass territory can take the form of a complex matrix, covering the three 'dimensions' of the bioenergy feasibility question:

- the **vertical** (or **techno-economic**) dimension, connecting the bioenergy market with: (i) biofeedstocks (agricultural, forest-derived, wastes, etc.), through (ii) conversion technologies (thermochemical, biochemical, etc.)

Figure 1.3 Bioheat to mellow the cold winters of the European North. General view of the multi-fuel heat generation plant in Sandviken, Sweden. (Courtesy of the company)

- the **horizontal** (or **techno-political**) dimension, linking the present large- and small-scale use of bioenergy to recent and ongoing R&D activities, as well as with the rest of the changing policy framework
- the *'third'* (or **socio-technical**) dimension, covering societal and space aspects, such as the regional variability on the European biomass scene, and the globalization vs localization balance of certain phenomena (e.g. 'think global, act local').

1.3.1 Present bioenergy contribution

The present contribution of bioenergy to the energy balances of EU countries is not very different from the 1995 data presented in Tables 1.4 and 1.4a.

We can see from these tables that the present energy contribution in the EU is mainly in the form of bioheat, whereas bioelectricity is usually limited to biomass co-firing in coal plants and waste utilization schemes, and transportation biofuels are used only in a few, rather experimental local situations. We should note that the European Commission's 'White Paper on Renewable Energy' envisages a doubling of current renewable energy utilization rates, from under 6% in 1995 to 12% by 2010 (based on total primary energy supply). According to these strategic targets, biomass should account for about 135 Mtoe/year in 2010 (with 75 Mtoe of bioheat); i.e. a three-times total increase (doubling of bioheat).

Market penetration rates of the three major bioenergy vectors are determined by the above defined three dimensions of feasibility. In the following parts of this section, the present situation will be summarized with respect to each of these groups of factors.

Table 1.4 Bioenergy in the EU energy systems (1995)

Country	Total primary energy supply (Mtoe)	Energy from biomass	% Bioenergy within all renewables	% Total electricity from biomass
Austria	27.0	11.1	49	4.2
Belgium	59.8	0.6	97	1.2
Denmark	25.4	5.9	93	6.3
Finland	29.9	19.1	79	18.2
France	245.2	4.0	66	1.4
Germany	350.5	1.3	71	1.2
Greece	27.4	3.3	64	0.7
Ireland	11.2	1.6	70	1.7
Italy	161.7	2.2	17	1.6
Luxembourg	3.5	1.1	86	0.1
Netherlands	85.3	0.0	94	2.1
Portugal	19.4	13.5	63	7.0
Spain	104.0	3.4	52	3.1
Sweden	52.7	16.7	66	12.5
United Kingdom	231.8	0.6	85	0.7
Total	1435	3.3	61	2.8

Sources: Eurostat; National Statistics; estimations.

Table 1.4a Biomass penetration in EU energy markets (1995)

Bioenergy vector	Contribution	Market penetration
Bioelectricity	64.8 TWh(e)/yr	2.8% of total electricity generation
Bioheat	38 Mtoe/yr	ca. 6% of total heat utilization
Liquid biofuels	< 1 Mtoe/yr	< 0.1% of total transportation fuel
Total bioenergy	44.3 Mtoe/yr	3.3% of total primary energy

Sources: Table 1.4; European Commission data; and author's estimates.

1.3.2 Techno-economic aspects

The present structure of the bioenergy techno-economic chains (see Section 1.2.3) is shown in Table 1.5. Only available feedstocks and technically mature technologies are listed.

Feedstock-related issues constitute a main factor for the low market penetration of biomass, determining the feasibility of all bioenergy vectors in two ways, i.e.

1   by increasing real biomass procurement costs, incorporating production, harvesting, transportation, handling and other such cost items
2   through biomass logistics affecting – besides costs – the seasonal availability, the transportation and storage requirements, and even the technical suitability (in the cases of varying composition feedstocks) of bioresources.

1.3.3   Techno-political aspects – research policy

No unified mapping of the biomass territory according to techno-political criteria is possible, as such 'maps' differ according to the particular policies examined in each case (see Table 1.3). As research-related questions and their effect on market penetration

Table 1.5 Present structure of the bioenergy techno-economic system

Energy vectors	Biomass feedstocks	Conversion technologies	Major constraints
Bioheat	Fuel wood Wood wastes Agro-residues Municipal and various wastes	Combustion Gasification Anaerobic digestion Landfill gas use	Feedstock logistics Feedstock cost
Bioelectricity	Fuel wood Wood wastes Agro-residues	Cofiring Combustion Gasification	Feedstock logistics Feedstock cost Suitability of new feedstocks
	Municipal and various wastes	Anaerobic digestion Landfill gas use	Technical improvements
Transportation biofuels	Sugar crops Starch crop Vegetable oils	Fermentation to bioethanol Oil esterification to bio-diesel	Feedstock logistics Feedstock cost Product logistics

Table 1.6 Present situation of bioenergy chains market proximity

Technical maturity /market proximity	Bioheat	Bioelectricity	Transportation biofuels
Research	Application-oriented	Gas cleaning Other technical improvements	New feedstocks Utilization of co-products Environmental aspects
Development	Application-oriented	Gas cleaning Other technical improvements New feedstocks	New feedstocks Utilization of co-products Engine tests Fuel quality
Demonstration	Testing new developments Optimizing logistics	Testing new feedstocks Testing new developments Testing new feedstocks Optimizing logistics	Testing technical developments Engine tests Local/regional experiments Search for strategic allies
Diffusion	Growth of existing markets Establishing new markets	Establishing markets Strategic partners search (utilities, cogen)	Limited by the above factors Acceptance by fuel companies not secure

will be of central significance according to EUREC's rationale, in Table 1.6 we have depicted the present bioenergy situation in the EU according to the level of technical maturity and/or market proximity of the biomass production chains already defined in Table 1.5.

The key words shown in each block of Table 1.6 comprise both actual research 'bottlenecks' and corresponding priority actions, e.g. as already included in the European Commission's 5th Framework Programme, as well as future priorities, e.g. for inclusion in subsequent, targeted research initiatives within the EU.

Through similar maps constructed with respect to other policy areas, such as agricultural, regional and environmental ones, we can identify their respective strategic bottlenecks and resulting priority actions. Application of these tools thus makes it possible to 'pave' strategic 'roadmaps' for all parts involved and to establish a common basis for policy co-ordination (see Section 1.2.5).

1.3.4   Socio-technical aspects

The 'third dimension' determining the feasibility of bioenergy schemes is, probably, the least known and is usually neglected in planning. Market penetration, especially of novel bioenergy technologies, is low, as European societies have shown little interest in the subject or are poorly informed.

In Table 1.7, we analyse the main aspects of the present acceptability of the main bioenergy vectors by societal bodies (but not governmental, this being part of the techno-political axis) in the EU, according to the level of action.

Table 1.7 Present situation of bioenergy vectors' societal acceptability

Level of localization	Bioheat	Bioelectricity	Transportation biofuels
International	Sustainabililty questions on forest management in developing countries	Not clear view	Favourable for 'green' fuels. Some ecological questions, e.g. on Brazil
European Union	Lack of specific support by EU bodies	Lack of specific support by EU bodies	Lack of specific support by EU bodies
Groups of states	Favourable in Nordic countries	Possible interest in the Mediterranean	Not clear view
Member states	Favourable attitude in several countries, e.g. DK, SE, FI, A Negative attitude for 'traditional' uses in other countries (GR)	Unfavourable attitudes where alternatives do exist (nuclear, hydropower)	Favourable in some countries, e.g. FR Possible reactions from eco-activists
Regions	Forested areas. Areas with high space-heating need. The case of district heating in Austrian areas	Areas with grid limitations (e.g. Crete) Islands Other isolated areas	Possible 'pilot' experiments in areas rich in feedstocks
Local communities	High variability. Possible local experiments	High variability. Possible local experiments	Limited local experiments due to fuel markets

From this analysis, we can conclude that:

- the European social acceptability situation is highly variable at almost all levels
- bioenergy options could face both positive and negative societal attitudes, sometimes within the same region or state
- this whole socio-technical arena is too important to be left outside planning, especially considering educational, information and other relevant actions.

Figure 1.4 Generating bioelectricity at the Greve-in-Chianti RDF (refuse derived fuel)-based plant near Florence, Italy. View of the boiler section. (Courtesy of the company)

1.4 GOALS FOR RESEARCH, DEVELOPMENT AND DEMONSTRATION

In light of the analysis in the previous sections, research[9] becomes a significant player on the bioenergy scene. At the same time, it is clear that a major re-orientation of the present RD&D activities towards specific goals, both updated and new, will be necessary. Such a strategic 'turn' should take into account:

- the EU and national policy targets, such as the 135 Mtoe/year mega-target for the year 2010 of the EC's 'White Paper'
- a thorough critique of the biomass research scene in Europe, North America and other parts of the world, from the point of view of covering the critical 'gaps' for large-scale market deployment of bioenergy
- data and trends from foresight activities in related fields, judged against earlier ones, such as the EC's *Cellule de Prospective* study in the early 1990s.[10]

In addition, the very concept of 'research' in the biomass field needs to be extended and updated to include non-technical, interdisciplinary and structural issues, and activities that don't strictly fall under the 'not-competitive' label.[11]

1.4.1 Feedstock-related research

As already noted, feedstock issues are to a great extent responsible for the high cost and low market penetration of all bioenergy vectors. Feedstock-related research should be oriented appropriately to minimize such barriers, as shown in Table 1.8.

Table 1.8 Research priorities for biomass feedstocks

Research goal	Research efforts (examples)
To minimize feedstock costs	Better study of the potential (databases, models, GIS tools, Internet applications, expert systems); optimal selection of energy crops; the agro-refinery option
To maximize feedstock suitability	Suitability maps; targeted agronomic studies; biotechnological improvements; development of standards and norms
To secure the supply of biomass	Efficient bioresource mapping; modelling the logistics of agro-residues and energy crops; pilot projects in various regions
To establish sustainable patterns	Matching resources to local ecology and culture; developing economic/environmental models as tools

1.4.2 Conversion-related research

In this category we find the goals for the most important technologies, both mature and emerging, as summarized in Table 1.9. We should note that the priorities proposed therein do not strictly serve the needs of a particular technology, as is the case with typical reviews (see Annex), but reflect the market forces and the need to overcome the existing penetration barriers.

1.4.3 Research on bioenergy end-uses

Research goals are grouped together here for both the end-user stage of the bioenergy chain (see section 1.2.3) and its interface with conversion technologies. The priorities of this new research area are shown in Table 1.10.

As recent experiences from the successful diffusion of various renewable energy applications in EU countries have shown,[12] bottom-up mechanisms play a critical role, and should therefore be given priority in the respective end-use related research efforts.

1.4.4 Research on bioenergy systems

Even if biomass feedstocks become plentiful and low-cost, their conversion technologies become highly efficient and environmentally clean, and the bioenergy vectors generated come closer to their users and markets, bioenergy chains and systems will still have to be put together from these components and be introduced within and/or along alien production chains and systems, e.g. in EU rural regions. The research goals listed in Table 1.11 attempt to define some priorities for the support of this critical, final step of system synthesis and materialization.

Table 1.9 Research priorities for biomass conversion technologies

Research goal	Research efforts (examples)
To minimize conversion cost of mature technologies (combustion, biogas, sugar/ starch fermentation, biodiesel)	Mass/energy efficiency improvements; novel reactor design; deeper view of the science involved; modelling process economics
To eliminate technical problems of emerging (advanced combustion, gasification) and promising (pyrolysis, bioconversion of cellulosics) technologies	Standardization of solid biofuels; better study of ash effects; systematic studies on argo-residues and energy crops; gas cleaning; bio-oil refining for fuels and chemicals; basic bioconversion studies; scale-up from lab to full-scale
To minimize the environmental impact of all types of biomass conversion systems	Air pollution-free combustion; fate of tars from gasification; liquid biofuel combustion studies; waste reuse and recycling; impact assessment and life-cycle studies
To enhance the market uptake of new biomass technologies	Diffusion studies; 'learning curves'; new decision-support tools; appropriate pilot applications; new entrepreneurial culture

Table 1.10 Research priorities for bioenergy end-uses

Research goal	Research efforts (examples)
To improve the technical interface of bioenergy vectors with their users	Demand-side energy technologies compatible with biomass; novel solutions, e.g. fuel cells; emphasis on quality of products and services
To improve the economic interface	Market surveys; novel financial and pricing instruments; between bioenergy/biofuels and their users innovative marketing strategies; role of retailing
To improve the environmental acceptability of bioenergy vectors	Labelling as a tool; use of environment-friendly materials and components; life-cycle analysis confirming 'green' image
To improve the social acceptability of bioenergy and biofuels	Effect of lifestyles and fashions; new approaches in biomass education, information campaigns and market strategies

Table 1.11 Research priorities for bioenergy systems

Research goal	Research efforts (examples)
To optimize the structure of the system according to the needs of its surroundings	Biosystems modelling; system dynamics; designing multiple food/non-food biomass option plants; local bio-refineries
To maximize the social and economic benefits for local/regional actors involved	Quantifying socio-economic externalities: employment, local income multipliers; economies of scope vs scale; training for new local/regional skills
To minimize environmental disturbances from the introduction of a new biosystem	Quantifying socio-ecological externalities; greenhouse gas balances; the biomass option for sustainable land-use planning
To maximize policy relevance and make use of policy synergies	Policy research; technological foresight, evaluation and assessment studies, Delphi workshops

Table 1.12 Outline of a European market-oriented biomass strategy

Strategy element	Targets for feedstocks	Targets for technologies	Targets for users and systems
Defensive	• Sustainable forest and wood management • Rational use of wastes and residues	Improvement of • Wood combustion efficiency • Feedstock quality (e.g. pellets) • Landfill gas systems	• Existing bioheat markets • Cogeneration • Solid biofuels – standards • Local systems
Aggressive	• Systematic use of bio-resources • Introduction of energy crops	• Supporting advanced combustion • Promoting gasification • Promoting biogas • Develop transport fuel from cellulosics	• New bioheat and bioelectricity uses • Cofiring at coal plants • Transport fuel additives • Regional systems
Exploratory	• Complex local/regional biomass systems	• New energy crops • Demonstrate pyrolysis • Develop hydrogen from biomass sources	• Fuel cells • Complex systems • Transport biofuels • New engines

1.5   A ROAD TO A STRONGER MARKET

Both present and potential markets for bioenergy look extremely fragmented. In addition to an optimal mix of the three types of energy vector (see above), a strong and healthy bioenergy market should have a place for both large- and small-scale applications, conventional along with novel solutions, centralized as well as decentralized schemes.

Outlining a possible European strategy that could turn that potential advantage into a real advantage, e.g. through appropriate synergies, is the subject of this last section of the chapter. A viable approach should consist of three different, interacting and co-ordinated strategic elements:

- a **defensive** element, focusing on the support of traditional bioenergy uses in rural European areas that are threatened by extinction
- an **aggressive** element, centred around the effort of penetration of new bioenergy technologies and product vectors in existing and new markets
- an **exploratory** element, oriented towards future biomass applications though the encouragement and support of innovations according to the lines set above.

The classification of characteristics of these three proposed strategic elements (Table 1.12) represents an effort to construct roadmaps for biomass in the EU based on the analysis reported in this document.[13]

ANNEX – BIOMASS CONVERSION TECHNOLOGIES[14]

This Annex gives an in-depth, strategic overview of the major technological pathways connecting biomass resources with their energy product markets. Six technological

Figure 1.5 Small, flexible bioenergy units, meeting the local feedstock/energy needs conditions, could be well integrated into rural development schemes. View of a rice-straw-based system in Indonesia. (Courtesy of Prof. A A C M Beenackers)

areas are reviewed, covering all established, emerging and promising cases. For methodological reasons, the format of each section follows closely that of the main text.

Combustion

Basic information

The majority of current biomass-derived energy comes from wood combustion. Combustion is the burning of biomass to produce heat, which can be used either directly for heating or drying or indirectly for producing steam to drive a turbine. This can be represented by the reaction

$$C_6H_{10}O_5 \text{ (biomass)} + 6O_2 \rightarrow 6CO_2 + 5H_2O + 17.5 \text{ MJ/kg}$$

Present state

Direct combustion processes for heat production and driving a steam cycle are already commercial, with a constant drive for the improvement of their efficiency and reduction of their pollutant emissions. Two types of boiler are commonly in use: those with fixed or travelling grates, and fluidized beds.[15] The former type is very common, ranging from the household boiler to large-scale 50 MW industrial furnaces, and can accommodate heterogeneous combustible materials. On the other hand, fluidized bed systems become attractive at plant sizes larger than 10 MW(th). Their main advantage is the ability to use mixtures of various types of biomass (woody, non-woody) and/or to cofire them with other fuels.

Current developments

A major development in this area is cofiring, i.e. co-combustion of biomass, mainly with coal, in existing, large thermal electricity plants, or wastes. New boiler concepts, where biomass is combusted together with coal, peat, RDF (refuse-derived fuel, i.e. upgraded urban waste fractions) or other fuels, can achieve high efficiencies, due to scale factors and reduced risks, as more than one fuel can be used, thus compensating for seasonal biofeedstock availability. In cofiring, an input of up to 10% of the calorific value of the fuel is presently targeted, but values up to 35% are envisaged. Biomass is chipped to small particles to make injection to the boiler possible. Other interesting cofiring options include the use of biomass pyrolysis or gasification (see below) as feedstock pretreatment stages. The latter technology is also foreseen to be implemented into existing gas-fired combined cycle plants, thus opening up an enormous potential for the introduction of biomass into existing energy systems.

Other promising areas of development include

* large combined heat and power (CHP) plants; at locations where heat has a significant economic value, stand-alone CHP units may become attractive
* highly efficient fluidized units can burn a mixture of fuels, even fuels containing up to 60% moisture
* the firing of biomass powder in ceramic gas turbines, expected to be commercialized in the years to come; these turbines, with a capacity of 100–500 kW, will generate heat and/or high-pressure steam, which can be used for power or CHP production.

Research goals

The main barrier is high costs, with the use of economies of scale. Most combustion R&D is on technical aspects, e.g. stoking, combustion air and fuel conveyance, aiming at improvements in combustion efficiency (>30%), reduction of pollutant emissions (e.g. fly ash), and the development of CHP plants. More research is also required on Stirling engines and pressurized combustion systems. Major R&D tasks lie in the field of cofiring: assessment of possibilities of co-combustion in different situations, development and demonstration of advanced boiler concepts. Specific topics to be studied include corrosion by alkalines and chlorides, slagging, as well as their prevention, thus contributing to the use of 'difficult' biomass types, such as straw, RDF and grasses. As all these developments will be helped if up-front investment is available, the involvement of industries will be an important issue. So, part of the R&D should concentrate on demonstrating the specific environmental and energy benefits of the new technologies to industries. Another issue to be assessed in future studies will be how well utilities meet the CO_2 and other emission standards.

Gasification

Basic information

Thermochemical gasification is the process of converting the organic part of biomass, at high temperatures, into a gas mixture with fuel value. This chemical reaction is endothermic, so heat has to be supplied externally. In the usual case where no air or oxygen is taken up into the process gas, a product gas of middle or high calorific value (10–18 MJ/Nm³) is produced, which makes downstream processing, e.g. gas cleaning, much cheaper. The other way to generate the heat required by the reaction is by partial combustion of the solid biofuel, according to an overall reaction, such as:

$$C_6H_{10}O_5 + 0.5O_2 \rightarrow 6CO + 5H_2 + 1.85 \text{ MJ/kg}$$

In the simplest systems, air is used, so the gas product is diluted with nitrogen, and gas calorific values are in the 3–8 MJ/Nm³ range. Although air gasification itself is relatively cheap, downstream gas cleaning is expensive due to the large volumes to be handled. On the other hand, the use of pure oxygen is expensive, particularly at small scale. Gasification is a rapidly emerging biomass conversion technology on the European and world scenes.

Present state
As in combustion, fluidized bed technologies are very attractive at large-scale (>5 MWth) gasification levels. A broad spectrum of biofuels can be gasified, but gas-cleaning costs will be high due to high tar production. Moving bed technology is an option at small scale; it is cheaper from an investment point of view, but requires the maintenance of a gas flow in the bed during an advanced state of conversion of the fuel. This means that substantial and costly biofuel pre-processing, e.g. pelletizing, is necessary. Counter current technology has high cold gas efficiencies but produces much tar, whereas co-current gasification is low in tar but has poorer cold gas efficiency (80% or less).

Current developments
Several biomass gasification processes have been and are being developed for electricity generation. In the BIG-ISTIG (biomass integrated gasifier-steam injected gas turbine), steam is recovered from exhaust heat and injected back into the gas turbine; in this way, more power can be generated from the turbine at higher electrical efficiency. The gas obtained by gasification can be combusted in a diesel, gas or 'dual fuel' engine, or in a gas turbine, after removal of tars, dusts and water. Its most effective and economical use is in the production of electricity via gas turbines combined with steam cycles. Yields higher than combustion can be obtained with internal combustion engines at lower power scales (50 kW–10 MW), whereas at higher levels (1–50 MW) combustion systems with steam turbines are more efficient. For very large power plants (50–100 MW), gasification can reach very high levels of efficiency through combined gas turbine–steam turbine systems.

Due to the potentially high efficiency and high temperature levels of the process, gasification is the main candidate technology for future large, biomass-fuelled CHP plants. The size and type of such plants depend on local conditions, such as fuel availability, composition and morphology, and the heat/power ratio required. Gas cleaning will be the critical step in this case too.

Research goals

Gasification research is aiming at large-scale (1000 t/d), oxygen- and/or air-blown systems, and high electricity production efficiencies, i.e. up to 42–47%. To reach this goal, emphasis will have to be given to simple and cheap gas cleaning technologies for dust, NO_x, NH_4, HCl, and alkaline components. This is needed for both large (>10 MW) and small plants. Stringent standards for fuel quality are also needed, as gasification is sensitive to changes in feedstock type, moisture and ash content, as well as particle size. More attention should be given to improving the tolerance of gasifiers to different types of biomass, and operating gas engines or gas turbines with low calorific value gases. Once efficient and cost-effective gasification has been achieved, the gas can also be used for the synthesis of secondary fuels, such as methanol or chemical feedstocks.

Pyrolysis

Basic information

Pyrolysis, a highly promising technology for future applications, is defined as the process of decomposition at elevated temperature (300–700°C) in the absence of oxygen. The products from biomass pyrolysis are solids (char, charcoal), liquids (pyrolysis oils), and a mix of combustible gases. The properties of each of these products depend on the reaction parameters. For centuries, pyrolysis has been practised for the production of charcoal (carbonization); this process runs at relatively low temperatures and high residence times to maximize solid char yield at around 35%.

Present state

In recent years, more attention has been paid to the production of pyrolysis oils, as they are easier to handle and of a much higher energy density than solid biomass. Oil yields of up to 80% by weight may be obtained by the process of fast or flash pyrolysis at moderate reaction temperatures, whereas slow pyrolysis will produce more charcoal (35–40%) than bio-oil. Pyrolysis liquids, currently referred to as bio-oils or bio-crudes, are intended for direct combustion in boilers, engines or turbines. Nevertheless, certain improvements by bio-oil upgrading are necessary to overcome unwanted properties, such as poor thermal stability and heating value, high viscosity, and corrosivity. Charcoal can be used in small gasifiers (of kW range) or may again become important as fuel. A main advantage of fast pyrolysis is that fuel production is separated from power generation.

Current developments

The flash pyrolysis process is still at a demonstration scale, whereas bio-oil upgrading processes are at a much lower development level. Pyrolysis can be of interest in conjunction with existing systems for large-scale electricity production, especially when biomass transportation is involved, as pyrolysis outputs are highly concentrated energy carriers, which can be used, for example, in existing coal boilers. Further, pyrolysis can be applied as a means to reduce the size of various waste streams, and recover valuable co-products. The largest plant built today is of 2 t/h, but plans for 4–6 t/h (equivalent to 6–10 MWe) are at an advanced stage. In the long term, pyrolysis oil could be used as transportation fuel (diesel substitute), after stabilization and other upgrading steps. This use is currently not economical, especially if no environmental credits are taken into account.

A number of technologies combining pyrolysis and gasification are also being developed in order to overcome the drawbacks of each technology and combine their respective advantages. Hydropyrolysis, a pyrolytic process under high hydrogen pressure, has actually more in common with gasification; it can also be regarded as a means to store hydrogen out of renewable resources, e.g. solar or wind, into a biomass-related carrier. A synthetic natural gas consisting of 80% of methane with similar combustion properties as normal natural gas can be made in this way.

Liquefaction is a low temperature (250–500°C), high pressure (up to 150 bar) process, in which a reducing gas – usually hydrogen – is added to the slurried bio-feedstock. The product is an oxygenated liquid with a heating value of 35–40 MJ/kg, compared with 20–25 MJ/kg for pyrolysis oils. The interest in liquefaction has been reduced due to economic and technical problems.

Research goals

On the R&D side – in addition to the above mentioned priorities – emphasis is to be given to the improvement of the production of pyrolysis oil for MW-scale power stations, and bio-oil upgrading using catalytic hydrotreatment; to solving the corrosivity and toxicity problems; to the modification of diesel engines, which will be run with pyrolysis oil; and to the development of pyrolytic oil-derived fine production of chemicals.

Esterification – biodiesel production

Basic information

Esterification is the chemical modification of vegetable oils into esters suitable for use in diesel engines: hence the term 'biodiesel'. Vegetable oils are produced from oil crops, e.g. rapeseed and sunflower, with the use of pressing and extraction techniques; the co-product is a protein-rich cake that is a valuable animal feedstuff. By the esterification reaction, plant glycerides in the presence of methanol or ethanol and a catalyst (usually aqueous NaOH or KOH) are converted to methyl or ethyl esters, respectively, with glycerine as co-product. The most promising biodiesel plant-oil ester is RME, i.e. rapeseed-oil methyl ester. The process scheme is summarized below (basis: 1 t RME production). Vegetable oil esters can be used in mixtures with diesel fuel up to 100%.

Extraction
Rape seed → Feed stuff + Rape oil
(300 kg)   (1900 kg)   (1000 kg)
Esterification
Rape oil →   Glycerine + RME
(1000 kg)   (110 kg)   (1000 kg)

Compared with conventional diesel, the use of biodiesel results in lower fuel emissions, and therefore can contribute to a reduction of respiratory problems and cancer risks. Due to the very low sulphur content of rapeseed oil, biodiesel combustion does not produce SO_2, which impairs lung function and contributes to acid rain. On the other hand, there could be some problems with odour – similar to that of cooking oil – when pure RME is used as a fuel.

Present state
The process for the production of RME is well developed and the product is commercially available in France, Germany and Italy. Recent results from the EC's Thermie Programme have shown no special problems, with conventional diesel engines working on mixtures of up to 50% RME.

At the same time, the resource basis for biodiesel production in the EU is rather limited. The total EU non-food oilseed production is confined to 0.7–1.2 Mha under the GATT Blair House Agreement (1993), and this allocation is being quickly taken up by member states, as shown by the growth of the EU actual cultivation areas: 0.2 Mha in 1993; 0.6 Mha in 1994. Most of this growth is due to rapeseed, which increased to *ca.* 0.4 Mha. The major producers are France and Germany, with 173 and 152 kha respectively (1994). On the other hand, biodiesel is expensive as a transport fuel, costing an extra €0.20–0.25/litre compared with its fossil-fuel equivalent. In the countries where RME is commercially available, it is competitive with fossil diesel only due to tax exemptions.[16]

Current developments
The main obstacles to the market deployment of biodiesel production technologies are the high production costs, the limited amount of raw material allowed to be grown in the EU, and the opposition of member states to biofuel detaxation. In a vicious circle, this lack of competitiveness of biodiesel in comparison with fossil fuels tends to discourage the injection of capital into the further development of esterification methods.

Research goals
To disentangle the esterification pathway from these obstacles, RTD should concentrate on testing RME in different types of engines; testing of engines for emissions (particulates) and reduction of odour problems; improved energy ratios and greenhouse gas benefits; and reduction of production costs, especially by a more efficient use of the process co-products (cake, glycerine).

Anaerobic digestion

Basic information
Anaerobic digestion is a biological process by which organic wastes are converted to biogas, i.e. a mixture of methane (40–75% v/v) and carbon dioxide. The process is

based on the breakdown of the organic macromolecules of biomass by naturally occurring bacterial populations. This bioconversion takes place in the absence of air ('anaerobic') in digesters, i.e. sealed containers, offering ideal conditions for the bacteria to ferment ('digest') the organic feedstock to biogas. A simplified stoichiometry of plant-biomass anaerobic digestion follows:

$$C_6H_{10}O_5 + H_2O \rightarrow 3CH_4 + 3CO_2$$

During anaerobic digestion, typically 30–60% of the input solids are converted to biogas. The co-products consist of an undigested residue (sludge) and various water-soluble substances.

Present state – current developments
Anaerobic digestion is a well-established technology for waste treatment. Biogas, either raw or usually after some enrichment in CH_4, can be used to generate heat and electricity through gas, diesel or 'dual fuel' engines, at capacities up to 10 MW(e). The average productivity is 0.2–0.3 m^3 biogas/kg dry solids. Nowadays, 80% of the industrialized world biogas production is from commercially exploited landfills. Further development of this technology as a biomass conversion pathway is rather limited.

Research goals
Current research is focusing on factors affecting optimal microbial population growth. High-solids digesters are being developed for the rapid treatment of large volumes of dilute effluents from various agro-industrial processes. This process has the advantage of a low-cost feedstock, and offers substantial environmental benefits as a waste management method.

Production of bioethanol

Basic information
This is another biochemical pathway, this time for the production of liquid, ethanol-based transport biofuels. In Europe they mainly have the form of an ethanol ether, ETBE (ethyl tertiary butyl ether), used as a lead substitute (up to 15%) in diesel/gasoline engines. In the USA, they have the form of gasohol,[17] i.e. a mixture of gasoline with ethanol, whereas in Brazil both pure ethanol and gasohol are used. The raw materials for bioethanol (biologically made ethanol) include sugar- and starch-rich crops, such as cereal grains, sugarbeet, potato, and sweet sorghum. Maize grain in the USA and sugarcane in Brazil are presently the most utilized worldwide biomass feedstocks for bioethanol production, whereas future feedstocks include various lignocellulosic plants and residues. Plant carbohydrates are converted to bioethanol by the following equation:

$$C_6H_{12}O_6 \rightarrow 2C_2H_5OH + 2CO_2$$

Present state

Bioethanol can be used as a pure fuel, as in Brazil's Proalcool Programme, or mixed with motor gasoline. If 100% bioethanol is used, engines should be adapted, while non-adapted engines can be used with mixtures. Ethanol can be also employed as fuel additives, i.e. a substitute for MTBE (methyl tertiary butyl ether),[18] and added to unleaded fuel to increase octane ratings. In Europe, the preferred ratios, recommended by the Association of European Automotive Manufacturers (AEAM), are 5% ethanol or 15% ETBE mixed with gasoline. ETBE is the product from the reaction of equal parts of ethanol and a hydrocarbon, isobutane; it has critical properties, such as octane rating, volatility, heat efficiency and corrosivity, superior to those of bioethanol. Gasoline–ETBE mixtures exhibit the same performance characteristics as gasoline, so engines do not have to be modified. ETBE can be manufactured in plants currently producing MTBE. The first industrial ETBE plant came on stream in 1990 at ELF France, using bioethanol supplied by the French producers Beghin-Say and Ethanol Union.

Current developments

The bioethanol production and distribution techniques are already commercial for sugar and starch substrates. The pretreatment and conversion technologies for the utilization of lignocellulosics are expected to be economical in the next 10 years, as some critical process stages are still at the pilot-plant level, while others need a strong R&D effort before commercial demonstration. From the point of view of economics, there is still a significant gap between fossil fuel prices (€0.15 /L) and that of bioethanol (€0.4–0.6/L) in Europe. This gap is expected to be reduced by improvements of industrial productivity and efficiency, based on the use of improved microbial systems, biotechnology and lignocellulosic feedstocks, as well as the utilization of co-products, e.g. lignin.

Research goals

A major obstacle to the development of bioethanol technologies is the lack of investment in RTD. In the absence of this investment, co-ordination activities should be initiated, also with the USA. Research on this topic is still in its infant phase: therefore an extensive list of topics will have to be addressed, including the development of advanced hydrolysis methods for lignocellulosic materials, efficient feedstock pretreatments, the use of novel yeasts, bacteria and fungi in fermentation, improved bioconversion schemes, identification of niche markets, and R&D in acetone-butanol fermentation, where a breakthrough is long overdue.

NOTES

1   A draft form of this document was discussed at the EUREC College of Members, Heiloo (NL), May 25-26 2000. Special thanks to Maria Luisa Delgado Medina, CIEMAT (ES); Herman den Uil, ECN (NL); Nicola M. Pearsall, University of Northumbria (UK); and Jacob Bugge, Folkecenter (DK) for their valuable comments. The significant contribution of Hubert Veringa of ECN (NL) to composing the special Annex on biomass technologies is also gratefully acknowledged.

2 This fiction should not be read in connection to any theory about the extraterrestrial origin of life and/or green plants. The use of extraterrestrials in this case simply serves as a cognitive tool, putting a distance between problems and potential solutions, in order to better identify the critical constraints for their successful matching.

3 Oakland, California, USA, September 1999.

4 '*Bioenergy*' should be a more exact title of this chapter, i.e. excluding all non-energy/non-food uses of bioresources; nevertheless, to avoid adding more confusion to an already 'fuzzy' subject, we stick to the widely used term 'biomass'.

5 Plus the risk of transgression of the professional 'turf' of one or more of the numerous distinguished experts in this vast field.

6 See reference 1.

7 As is usual with most biomass overviews, focusing either on conversion technologies or on feedstocks.

8 See the 'magnificent gift' above (Section 1.1).

9 In the following, unless otherwise specified, the term 'research' will be used to denote the whole research–development–demonstration–dissemination sequence of innovation phenomena.

10 See also on EU's biomass strategy: E G Koukios, D Wright, *Biofutur*, June 1992.

11 Of particular attention within this extended research agenda is the selection for support of the most promising pilot/demonstration/diffusion actions, acting as the 'locomotive' of the whole RD&D 'train'.

12 Major examples: biomass district heating in Austria; wind energy in Denmark; domestic solar heaters in Greece.

13 Table 1.12 can also be viewed as the European response in-a-nutshell to the 'magnificent gift' offered by the 'extraterrestrials' (see Section 1.1), thus bringing the quest of this chapter to a 'happy ending'.

14 This Annex is based on information from EUREC's earlier Position Paper (1996), as revised and updated by Hubert Veringa of ECN (NL), with final editing leading to its present form by E G Koukios.

15 In a fluidized bed, the constantly injected combustible particles together with the granular bed material are carried by a constant flow of gas in an upward direction. The bed constitutes the major heat capacity of the system and thereby stabilizes the process. In this way, effective heat and mass transfer are being taken care of. Such a system can combust a wide range of materials including fuels of non-biological origin.

16 The Schrivener Directive (1995), which is still under debate, proposes that EU biofuels be made economically competitive through a reduction in excise tax.

17 Gasohol is the term used in the United States to describe a mixture of gasoline (90%) and maize-based ethanol (10%); it should not be confused with gasoil, an oil product used to fuel diesel engines.

18 MTBE, obtained from fossil methanol, is added to unleaded fuel to increase its octane rating. Its output is currently growing at 10% annually in France. MTBE is the principal competitor of bioethanol and ETBE as an octane booster, with more than 10 Mt/year produced worldwide. Its price is linked to that of methanol, which exhibited major price fluctuations in the area of 95–190 ECU/t between 1987 and 1992. As a result, manufacturers may favour ETBE, whose market price is generally more stable. European regulations currently specify maximum MTBE (and ETBE) content in engines as 10% (v/v).

BIBLIOGRAPHY

Ballaire, P, *Publication Interne*. ADEME, France, 1996.

Biomass Technologies in Austria: Market Study. Thermie Programme, Action BM62, European Commission DGXVII, July 1995.

Energy in Sweden. NUTEK (Swedish National Board for Industrial and Technical Development), 1994.

Mauguin, Bonfils and Nacfaire (eds), *Biofuels in Europe: Developments, Applications and Perspectives 1994–2004*. Proceedings 1st European Forum on Motor Biofuels, Tours, 1994.

Proceedings of Biomass Conferences of the Americas: 1993 (Burlington), 1995 (Portland), 1997 (Montreal), and 2000 (Oakland).

Proceedings of European Biomass Conferences: 1989 (Lisbon), 1991 (Athens), 1992 (Florence), 1994 (Vienna), 1996 (Copenhagen), 1998 (Wuerzburg), and 2000 (Seville).

Wrixon, G T, Rooney, A and Palz, W, *Renewable Energy 2000*. Springer, 1993.

2. FUTURE RESEARCH AND DEVELOPMENT IN PHOTOVOLTAICS
Outlook and roadmap

H A Ossenbrink (European Commission, Joint Research Centre)

2.1 STRATEGIC SUMMARY

Photovoltaic (PV) technology for the conversion of sunlight into electricity is already a cost-effective method for many applications worldwide. This is due largely to more than 25 years of continued research and development, which has led to rapid market growth in the last decade. If PV is to displace a significant proportion of conventional electricity generation, at least another 25 years of aggressive market growth is required. This growth will rely both on the continuous improvement of well-proven technology, and on the introduction of new technologies. To achieve the ultimate cost reduction, substantial R&D efforts are required. The 'roadmap' for photovoltaic research and development assumes continuous market growth, stimulated by market introduction programmes and legislation that allows fair access to the existing electricity grid and gives PV support similar to that given to traditional energy technologies. Increased production will allow industry to refine process and manufacturing technology, and allow research groups to focus on new concepts that have a quantifiable time-to-market.

This chapter describes the measures that need to be implemented urgently in order to overcome imminent limitations on the availability of materials. Thin-film technology has yet to show that it can be scaled up to commercial levels, and efforts in this direction need strongest support.

Ultimately, the success of photovoltaic solar electricity still depends on the need to develop advanced storage technologies. This chapter calls for additional research so that storage solutions developed for other, mobile applications can be adapted for use in photovoltaic installations.

This chapter also makes the point that the pressure on research groups to stay close to markets and products often reduces funding for the search for new solutions in the mid and long term. Alternatives to the low-cost and/or highly efficient solar cell must be pursued if Europe wants to reaffirm its leading role in developing what may be the key industry of the 21st century.

2.2 INTRODUCTION: PHOTOVOLTAICS IN A NUTSHELL

Photovoltaic solar cells convert sunlight directly into electricity, without the need for intermediate thermal processes. Progress of the transfer from basic principles into a workable technology was strongly influenced by the development of semiconductors in the 1950s, and the need to provide electricity for satellites in earth orbit.

As the photovoltaic effect has so far worked best with semiconductor materials, for which the production cost per m^2 has decreased by a factor of 1000 since the 1970s, solar cells can also be made available for terrestrial applications. Today, countless off-grid applications are powered by photovoltaic generators, most of which are cost-effective when compared with conventional electricity generation. More recently, concerns about sustainable electricity supply have led to market introduction programmes aimed at increasing the share of renewable electricity generation through grid-connected applications.

In 2000 the worldwide annual production of photovoltaic generators reached 280 MW, equivalent to an area of some 3 million m^2. Each second, at least three solar cells leave a factory production line somewhere in the world.

2.2.1 Principle of operation

There are two basic methods for converting solar radiation into electricity for terrestrial applications: thermal and photovoltaic. **Thermal conversion** uses the energy of photons received from the sun to generate heat, and then uses conventional technology to convert this heat to electricity. **Photovoltaic conversion** attempts to transform all the energy of the photons into electricity directly by taking advantage of the intrinsic photovoltaic effect, which can be best realized in layers of modern semiconductor materials. A diode is formed when two layers of semiconductor are doped (by the addition of a donor or acceptor material) so that one conducts negative charge carriers and the other

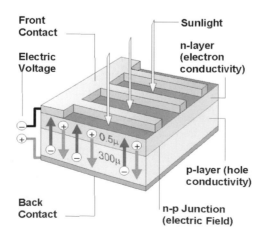

Figure 2.1 Principle of operation of a solar cell.

positive charge carriers. When photons fall on these layers they transfer energy to the charge carriers in an amount that depends only on the diode's material properties. The electric field across the junction of the layers separates the photo-generated positive charge carriers (holes) from their negative counterparts (electrons) so that electrical current can be collected at the front or back contact respectively.

2.2.1.1 Efficiency

Photovoltaics could ideally achieve about 85% conversion if all the energy of the absorbed photons was transferred into that of charge carriers. This is the aim for the new 'third generation' approaches that are under development, but current cells have a conversion limit of about 40%. This is because any transfer of energy from photon to charge carriers can only be of the amount given by the band-gap of the semiconductor material. Photons as received from the sun have energies between 0.2 and 5 eV. Photons with energies below the band-gap energy (which is 1.12 eV for silicon, for example) do not create charged carriers, being lost instead to thermal excitation and transmission. Photons with energies above the band-gap transfer only the band-gap energy; any excess energy is lost for the photovoltaic effect. It might be possible to improve efficiency by using photovoltaic solar cells that consisted of multiple layers with different band-gaps; each photon would be absorbed in the layer where its energy matched the band-gap.

2.2.2 Production technology

The following description is restricted to crystalline solar cells, which in 2000 accounted for 87% of the world's production.

2.2.2.1 Ingot

The starting material for ingot production is metallurgical grade (MG) silicon, which is obtained by chemically reducing SiO_2. MG silicon is available in large quantities, as it is an important base material not only for the electronics industry but also for the chemical industry. A small proportion of MG silicon is purified to electronic grade (EG) for the microelectronic industry, from which the PV industry receives off-specification material. MG silicon has too many impurities to yield an efficient solar cell material. Almost all ingot formation for photovoltaic purposes is based on one of two distinct crystallization and purification processes: Czochralski (Cz) growth for single-crystal silicon or directional solidification for the multi-crystalline material.

2.2.2.2 Czochralski growth

EG silicon is melted in a crucible, which is rotated, and a seed crystal is dipped into the melt. Silicon crystallizes on the seed at the surface of the melt. By adjusting the pulling and rotation speeds, single-crystal silicon ingots can be grown up to 1 m long with diameters of more than 20 cm and weighing up to 50 kg. Impurities remain in the melt. The crystal has the same uniform lattice structure and orientation as those of the seed crystal.

2.2.2.3 Casting and directional solidification, leading to multicrystalline ingots

EG silicon is first melted in a crucible, followed by controlled cooling from the bottom of the crucible. When the silicon reaches its solidification temperature, it rapidly transforms into its solid state. Ingots can be as large as 60 cm × 60 cm × 30 cm, and weigh up to 300 kg. Impurities segregate at the boundaries of crystalline grains and in the top of the section of the ingot. Crystal orientation is random between grains. Certain impurities located in the grain boundaries and inside the grains can be passivated in later cell-processing steps.

2.2.3 Wafers and junction formation

The ingots are cut into wafers approximately 0.3 mm thick using wire-saws, which allow up to 400 slices to be cut in one sawing process. The kerf loss amounts to about 50% of the ingot material. Surface damage caused by the sawing process is treated by chemical etching. The junction of the wafers is created in a diffusion furnace: either phosphor or boron is diffused into the wafer to create either n- or p-type layers respectively. As the diffusion occurs everywhere on the wafer, the edges have to be etched by a plasma-etching process, and the rear side has to be compensated with the right dopant.

2.2.4 Cells

To finish the solar cell the front and back contacts have to be created on the wafers. They are usually deposited by screen-printing conductive paste, which is then sintered to ensure good contact to the surface of the silicon. Some manufacturers prefer a metallization process, in which metals are electrolytically deposited, sputtered or

Figure 2.2 A typical photovoltaic module during mounting on a residential roof. This is a standard module, with aluminium frame, tedlar back-sheet, and 10 × 10 cm² mono-crystalline cells. Power is about 50 W, current 3 A, voltage 17 V.

evaporated through a mask that defines the pattern of the contact grid. Tabs consisting of solder-coated copper strips are soldered to the front and back main contacts to collect the current. An anti-reflective coating of titanium dioxide (TiO_2) is usually applied to the front surface of the cell, to reduces losses caused by reflections of sunlight. Some manufacturers are now using silicon nitride (Si_3N_4) instead of TiO_2.

2.2.5 Modules

The solar cells are connected in series by soldering the front tabs of each cell to the rear side of the next cell in order to yield higher voltages as required by the application. These strings of cells are embedded in an encapsulant (EVA, ethylene vinyl acetate), which provides a better optical interface between the cells and the air, and also seals the solar cells against humidity penetration at the edge of the module. The front of the module usually consists of toughened glass, and the rear of tedlar polymer foils. The compound is laminated under elevated temperature and vacuum. If required, a metal frame is applied to the finished laminate to provide increased stability and additional mounting points.

2.2.6 Photovoltaic systems components

Photovoltaic systems today consist essentially of the following components.

The **photovoltaic module** encapsulates between 10 and 350 single solar cells. The minimum number of cells is dictated by the voltage that the module should deliver. Typically the voltages correspond to the charging voltages for lead–acid accumulators (32–36 or 72 cells in series for 12 V or 24 V accumulators respectively), but modules with a much higher number of cells (up to 350) are made, in particular for the building market. The total number of modules depends on the power needed and on the overall configuration of the system in terms of voltage and current.

An **electricity storage accumulator** is provided if required, together with an electronic charge regulator. For most accumulator technologies, protection circuits against overcharging or deep discharge increase the reliability of the system.

If alternating current (AC) is required at the usual 230 V, 50 Hz standard (or 110 V/60 Hz), an **inverter** is necessary to convert from the direct current (DC) delivered by the PV modules. These inverters are either autonomous, for accumulator systems, or designed to be connected to an existing AC distribution grid. In these applications no storage accumulator is necessary. The existing electricity grid either delivers the power not generated by solar energy (e.g. at night) or absorbs the surplus production.

Often the **load** to be powered by PV requires additional electronic converters for improved matching to the current and voltage characteristics of the photovoltaic array. This is particularly the case when water pumps or resistive loads are to be powered by photovoltaics. These **maximum power point trackers** adjust the operating point of the load so that the maximum power point of the photovoltaic array is always maintained, regardless of solar radiation and load resistance.

2.2.7 Applications

Photovoltaic electricity generation is characterized by a low energy density. Unlike other, conventional energy sources, the efficiency of PV is not greatly influenced by size, and so photovoltaics can be applied in small, decentralized plants as well as in large, central power plants. Moreover, photovoltaic systems can be built in sizes from centimetres to kilometres and can be deployed at any location, often with modularity allowing extensions. These particular features make many applications possible. For example:

- power up to 5 W: pocket calculators, small chargers, radios, remote wireless sensors
- power up to 500 W: small lighting systems, call-boxes, traffic signals, navigation lights, small communication systems, solar home systems, medical refrigeration, cathodic protection
- power up to 50 kW: pumping systems, desalination, propulsion of recreation boats, stand-alone supplies for isolated buildings, rooftop systems (grid-connected)
- power up to 500 kW or more: large, grid-connected systems, either integrated into a building (roofs, facades) or stand-alone.

Even though only a few larger applications have been realized so far, there is no upper limit for a PV power plant other than the available area. If the PV generator could be installed on floating pontoons in the ocean, or in a desert area, even this limitation would not exist.

2.2.8 Potential and features

Photovoltaic electricity generation is a mature technology. It can provide its services virtually anywhere in the world, to any application. Some of its main advantages are as follows:

- It is non-exhaustible, and no materials are consumed in generating electricity.
- There is a sustainable, unlimited supply of the primary raw material (sand).
- There is no pollution or waste in operation, and only a small amount of pollution during production.
- It can produce electricity anywhere on the planet.
- Its modularity makes it suitable for decentralized generation.
- It is extremely reliable.
- It is compatible with human activity.

It is due to these advantages that since 1985 the photovoltaic industry has been able to increase sales by 15–20% per year, and recently even up to 25–40% per year. Many applications are cost-effective, either because there is no alternative, or because competing technologies would be more expensive.

Consequently, the largest cost-effective application of today is the provision of electricity to rural communities in the developing world and emerging countries. The current realistic potential, which would include electrification not only of Latin America and Africa but also of India and Southeast Asia, is estimated to be ultimately larger than 500 GW$_p$ – more than 1000 times the current production.

2.2.9 Current cost and cost reduction potential

2.2.9.1 Stand-alone systems

Typical small electricity systems, including storage and balance of system (BOS) components but without appliances, are sold for €8–10/W$_p$ for a 50 W$_p$ system (system price less than €500). However, it would be misleading to calculate only the energy cost for such a system. Based on present financing concepts it is more appropriate to list the services that can be provided by photovoltaics for the equivalent price of €1 (see Table 2.1).

Table 2.1 Service obtained for the price of €1 using PV-powered systems

Service	Quantity	Usage	Power
Light for a single family	14 days	4 h/day	15 W
Television	7 days	2 h/day	50 W
Food refrigeration	2 days	12 h/day	25 W
High-quality water (rev. osmosis)	30 litres	Rev. osmosis	

GRID-CONNECTED SYSTEMS

In mid-Europe, grid-connected systems of the 3 kW size produce electricity at about €0.40–0.60/kWh, and in southern Europe at about €0.25–0.35/kWh. These electricity costs are based on a system price of €4–5/W$_p$ and a lifetime of 25 years. These systems are usually owned by the user of the electricity, and so one can assume that the cost equals the price.

For the next 10 years, photovoltaics will produce electricity at costs that are still considerably higher than that of the bulk electricity production of European electricity grids. However, we can expect costs to decrease rapidly, for the following reasons:

• Prices for PV installations are currently decreasing by 5–10% per year, as a result of competition, improved technology, the building-up of large (100 MW$_p$) production facilities, and the results of the learning curve. However, as PV prices are usually quoted in US$, exchange rate mechanisms could change this factor up or down.
• If the current (1998–2000) average production growth rate of 25% per year is maintained, in 2010 the worldwide production will be 10 times that of today. From the learning curve, this scale-up alone will allow production at half the costs.
• Technology improvements will be implemented in the near future at low capital costs: manufacturing technology will be refined to increase throughput at high yield, and there will be increased use of modified equipment originally developed for the

electronics industry and now considered to be obsolescent. This applies in particular to the processing of larger, thinner wafers: PV is currently using wafers in sizes up to 6 in (150 mm), whereas equipment for electronic components is already processing wafers of 8–12 in (200–300 mm). It also applies to the use of systems originally developed for thin-film depositions of display systems. In the long term, however, the cost-effective production of PV requires dedicated manufacturing equipment, not equipment originally intended for the semiconductor industry. Fortunately, some equipment manufacturers have already entered the PV field.

2.3 ACHIEVEMENTS

2.3.1 Current situation

2.3.1.1 Worldwide production quantity increases to 288 MWp

From 1995 to 2000, worldwide shipment of photovoltaic modules increased from 80 to 288 MW_p, an average increase of 27% per year. The increase of shipment in Japan in 2000, at almost 50% per year, and in India, at 40% per year, is worthy of note. Today, Japan supplies nearly half the world PV market. In 2000, thin-film solar cell production accounted for about 12% of production. 85% of production technology was evenly distributed between mono-crystalline and multi-crystalline silicon technologies. Other crystalline technologies, such as ribbon growth and thin silicon films, contributed slightly more than 3%.

2.3.1.2 Investment in manufacturing plants

During the period 1998–2000 almost all PV producers invested in the expansion of production facilities. The steady growth rate resulting from new market initiatives and the developments in rural electrification have allowed companies to speculate on a return on investment in the near future. In addition, new start-up companies in Europe and the USA have increased the world's production capacity. In 2000, solar cell production capacity is estimated by the authors to be approximately 300 MW_p per year. Given the growth rates of the past two years, this capacity will be fully used by the end of 2001 at the latest.

2.3.1.3 CIS (copper-indium-diselenide) modules in the market

In 1999, the first CIS modules were available through retailers in Germany and the US. Sale prices were below €8/W_p for a single 40 W_p module with 9% efficiency. Also, a new company in Germany is currently constructing a production line for CIS modules.

2.3.1.4 Construction of the first factories for CdTe module production

After years of research, CdTe technologies seemed mature enough for a German company to invest in building a factory with an annual capacity of 10 MW_p. The first modules were expected to be shipped in 2001.

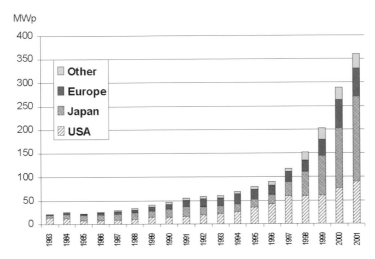

Figure 2.3 Worldwide sales of photovoltaic modules, including small consumer applications (excludes modules made from purchased cells). (Source: P Maycock, PV Insiders Report)

2.3.1.5 High-volume grid-connected systems

In the past five years, a number of regions in the world have launched programmes for small residential grid-connected systems:

- In 1995 Germany launched the 1000 Roof Programme, which installed 2250 systems of about 6 MW_p in total.
- Japan launched the 70,000 Roof Programme, to be realized over a period of 10 years.
- The Netherlands supports a 250 MW_p programme for residential PV applications.
- Italy started the 10,000 PV Systems programme.
- The United States started a 1 Million Solar Roof Initiative.
- The European Commission, after presenting a *White Book on Renewable Energy Sources*, launched the Campaign for Take-Off, aiming to install 300,000 building-integrated systems.
- Germany is subsidizing 100,000 roofs by providing low-interest loans.
- Spain allows payback rates of €0.40 /kWh.
- Germany has introduced a law to allow payback rates of up to €0.50/kWh for grid-connected photovoltaic installations.
- Belgium (Flanders) has introduced an investment subsidy of 50% for grid-connected, building-integrated PV installations up to 3 kWp, with an additional subsidy of 25% from the utility.

These initiatives alone will represent an approximate power of 2000–3000 MW_p, when installed.

2.3.1.6 Reliable and mature inverter technology

Since the residential rooftop programmes were launched in Germany and Japan, considerable improvements have been achieved for grid-feeding inverters. The progress was not so much in efficiency (already steady at an excellent 90–94%), but more in reliability and experience in mass production, particularly in the 1–5 kW power range. With increased production volume and the availability of new power semiconductors, prices continue to fall. This also influences the market for >50 kW inverters as, increasingly, string inverters are being chosen over single, large inverters. String inverters match the substrings of an array; such installations have a high number of small inverters connected to the grid. The progress of inverters integrated into single modules, which could ultimately be the solution for all small grid-connected systems, should also be noted.

2.3.1.7 Worldwide efforts to ensure quality of small, stand-alone PV systems

Thanks to the increasing efforts of international organizations, PV applications for rural electrification currently show strong growth rates, of similar magnitude to those of residential systems. To ensure user satisfaction and quality of such systems, US, European and Japanese industry associations have teamed up to create PV-GAP (the Photovoltaic Global Approval Program). PV-GAP is involved in fostering quality standards, the certification of PV systems, manufacturers and installers, and worldwide training programmes.

2.3.2 Updated goals

Table 2.2 summarizes and compares the current achievements with the goals set out in the 1995 EUREC Position Paper.

The 1995 EUREC Position Paper set goals for the years 2000 and 2010, which are reported here. This work is setting a new goal for five years from now (2005), and maintains the EUREC 1995 medium-term goal for 2010.

2.3.3 Main barriers to overcome

Photovoltaic solar energy conversion is still the most expensive form of renewable energy today. However, it has the largest potential, and can easily be integrated into current electricity systems. The main barriers to overcome are as follows.

2.3.3.1 Large-scale market introduction

The main barrier to overcome is the high price. The price needs to be further reduced by a factor of four or five for photovoltaic electricity to become competitive with bulk, conventional electricity production. Scale-up of processes, driven by large market introduction programmes, will be the major cost-reduction factor for the next few years, on the condition that material is available at economic cost.

Table 2.2 Status, achievements and goals compared with the 1995 EUREC Position Paper

	Status 1995	Goal for 2000	**Achieved 2000**	Short-term goal for 2005	EUREC 1995 medium-term goal for 2010
Solar cells					
c-silicon					
laboratory	23%	24%	**24.7%**	25%	26%
production	12–15%	16–18%	**15–17%**	16–18%	>20%
mono-crystalline			**13–15%**	14–16%	
multi-crystalline					
Thin-film (polycrystalline, amorphous)					
Laboratory	10–12%	18%	**18.8%**	20%	20%
Production	5–8%	>10%	**10%**	12%	>15%
Advanced devices (tandem, concentrator)					
Laboratory	25–27%	>30%	**33%**	34%	35%
Production	18–19%	>20%	**24%**	25%	25%
			(GaAs stacks)		
Modules					
Efficiency	11–13%	–	**13%**	15%	
Lifetime		>20 yr	**25 yr**	30 yr	>30 yr
Degradation during lifetime		<10%	**<5%**	<3%	<10%
Cost €/W$_p$		<2.5	**<4.0**	<2.5	<1.5
Of which cell cost	>3.0	<1.5	**<2.5**	<2.0	<1.5
Inverter					
Efficiency					
100% load	95%	>97%	**96%**	97%	>98%
10% load	85%	>90%	**90%**	92%	>95%
Cost €/W$_p$					
<500 W	1.5		**1.2**	0.6	
1 kW–5 kW	1.0		**0.5**	0.3	0.3
>5 kW	0.8		**0.6**	0.3	
Accumulator					
Efficiency					
Ah	83–90%		**85%**	85%	
Wh	76–82%		**80%**	80%	
Cost €/W$_p$					
Investment €/kWh$_{capacity}$	160–200		**80–120**	65–100	
Specific energy €/kWh$_{from\ battery}$	0.11–0.33		**0.20–0.30**	0.15–0.25	
System cost €/W$_p$					
Stand-alone		<5.0	**<8.0**	<5.0	<3.0
Grid connected 1–3 kW	8–12	<4.0	**4.0–5.0**	<3.5	

2.3.3.2 Crystalline solar cells

The availability of an economically priced silicon feedstock will be a major barrier for further cost reduction if it is not overcome soon. In parallel, better use of material has to be made by alternative production processes. For the conventional technology, more progress has to be made to further reduce the processing steps and handling requirements, in order to improve manufacturing throughput, increase yield from silicon to module, and make better use of the expensive equipment.

2.3.3.3 Thin-film technologies

The main barriers that need to be broken through for these technologies to succeed are:

*   the bad image of the first, amorphous-silicon prototypes
*   the relatively low efficiency
*   the toxicity of materials in production and disposal (more a perceived than a real, proven problem)
*   the lower production yield.

All four problems have to be addressed simultaneously if thin-film technology is to take a major share of the growing market.

2.3.3.4 Electricity storage

New battery technologies that are being developed for other applications need to be adapted for photovoltaic use, even though at present they appear to be more expensive than current lead–acid accumulators. The barrier to overcome is to advertise the reliability, maintenance and ease of use of stand-alone applications. It is also important to look more closely at applications that do not need electricity storage, because the actual product can be stored, such as water pumping and desalination, ice production, and propulsion. Long-term research needs to focus on electrochemical processes that can both convert and store electricity.

2.3.3.5 Capital costs and investment payback

All investment in production facilities has to be returned through profits from products sold. In terms of investment appraisal, the equipment costs of production processes have to be assessed, whether those processes are already established or are new implementations of R&D. Cost targets should be formulated not as cost per watt manufactured, or per kWh delivered, but as the ratio of investment cost to throughput and yield of a manufacturing line. This would bring estimates of cost reduction much closer to the R&D process. It would also encourage the development of processes that could benefit from used or obsolete equipment from other, non-photovoltaic manufacturing. Clearly, industry has to be much more closely involved from the beginning of this investment appraisal process.

2.3.3.6 Neglecting the future

For maximum cost reduction new technologies will be needed, as there are probably limits on what can be achieved with current technologies. In the race to lower cost and assess each R&D proposal for its immediate technology transfer potential, resources are taken away from a forward-looking, often indirect approach to exploit new avenues of photovoltaic conversion. The quest for clean electricity for the future is too important to leave breakthroughs to 'surprise' technologies, which might or might not appear. Other research areas, such as molecular biology or electronics, have a much more formal approach to forward-looking research and cross-fertilization.

2.4 ROADS TO A STRONGER MARKET

Market development, even though not technical, is essential for further growth. It is beyond the scope of this chapter to list all possible actions for market growth, but the following description addresses the three main roads to be followed in order to ensure the steady growth of photovoltaic applications.

2.4.1 Setting the legal, fiscal and political framework

The future development of the industrialized countries must be sustainable. On the political agenda of most of these countries is the introduction of renewable energies, in order to meet internationally agreed CO_2 abatement targets. However, a common approach to common problems is often lacking, and depends on the environmental awareness of the citizens. The European Union has endorsed in the *White Book on Renewable Energy Sources* a general policy to introduce renewables on a larger scale, and has set as a target the doubling of the amount of renewables by 2010 compared with 1998. Similar policies have to be established in other countries to set the political framework for legal, fiscal and promotional action.

A market introduction scheme is required that ensures fair access to the electricity grid by compensating for the still higher cost of this particular renewable energy technology. If this compensation, which reflects the external costs of conventional energy use, is not to be paid from public subsidies, it should be distributed among all users of electricity. It is important that such schemes are introduced transnationally, to avoid market distortion.

The legal framework needs to define the conditions for accessing the existing grid, the technical standards for quality, reliability and safety, and the relation to the operators of the grid. The fiscal measures need to support the introduction by means of (income) tax credits and exemption from VAT or eco-taxes, as far as is compatible with national or international trade agreements.

Applications of photovoltaics for rural electrification need new policy frameworks in development policies, and need to benefit from fiscal measures such as exemption from import tax. This tax, together with lack of training, is the strongest obstacle to the widespread introduction of photovoltaics in developing countries.

2.4.2 Advertising the benefits

The benefits that photovoltaics can bring to professional applications such as communications transmitters and navigational aids need to be better and more widely advertised, to emphasize the added value that they can provide in terms of quality, reliability and delivered service not only for portable devices, but also in particular for services such as water pumping, desalination and upgrading. The cost of these services and products made available through photovoltaic electricity should be advertised in relation to the market value, rather than to the price of electricity. Advertising should stress such advantages as the broad availability everywhere on the planet, as this is important in areas with difficult access or where unattended operation is required.

2.4.3 Building the 21st century image

It is time to create a new image for photovoltaics, using professional advertising campaigns. The focus should be on building a 21st century image: high-tech but green, available everywhere, giving mobility to the user, being integrated to create new looks, allowing entirely new applications, an overall new product with new benefits. The people currently involved in photovoltaics are not the real experts in image building, as they tend to highlight the technical features. There are good examples in the advertising of cars, mobile communications and even electricity providers that show the way to go.

2.5 GOALS FOR RESEARCH, DEVELOPMENT AND DEMONSTRATION (RD&D)

2.5.1 The way forward: the R&D roadmap

Photovoltaic solar energy conversion is the most expensive form of renewable energy today. However, it has the largest potential, and can be easily integrated into existing electricity systems. There are four components that constitute the price of photovoltaic electricity: the available solar resource, the price per unit of power, the capital and O&M costs, and the lifetime. Consequently, the main issues to address are:

- the high manufacturing costs
- the development of dedicated manufacturing equipment
- the difficulty of scaling up deposition processes for thin films
- materials availability, and bottlenecks in the production chain
- reliable and cost-efficient system components
- standardization and quality control
- increased emphasis on R&D for third-generation solar cells.

Many national and international research programmes in photovoltaics set out goals to achieve a certain cost target within a certain time. Future R&D roadmaps (including this one) need new approaches to achieve the final goal, which is to produce electricity by means of photovoltaics at competitive costs.

Cost reduction can be achieved by increasing the production capacity of a single manufacturing plant. Comparison with similar technologies suggests that a cost reduction of a factor of 2 can be expected whenever production increases by a factor of 10.

These economic scale-up laws are based on **experience curves**, which relate the cost per item to its cumulative production quantity. At regular intervals, when a technology has been optimized to achieve a cost minimum, new – initially more expensive – technologies appear. After breaking even, these then take over the bulk of production. Perhaps photovoltaic technologies are close to such a point, where thin-film technologies could take over from crystalline silicon technologies.

2.5.2 Crystalline silicon solar cells

Solar cells produced from crystalline silicon today constitute about 85% of the world's module shipments. In terms of electricity production, the share is higher. Mono-crystalline Czochralski material and multi-crystalline (cast) each contributes about 42% to the market; the rest comprises other crystalline technologies such as silicon thin films or ribbon cells. Single crystalline ingots are pulled from molten silicon, yielding a homogeneous crystalline structure. When molten silicon is solidified in a controlled cooling process, the material obtained is multi-crystalline. The crystal orientation of the grains is random and gives these solar cells their characteristic appearance (Figure 2.4).

Silicon wafers can also be grown directly from the melt by, for example, ribbon growth or direct casting, thus eliminating kerf losses. Currently the only large-scale commercial process of today is the **edge defined film growth** (EFG) process of RWE Solar (formerly TESSAG). Hollow octagons are grown from a silicon melt up to 5 m long, and a laser is used to cut single cells from these tubes. Almost 90% of the Si ingot is used for the solar cells, which have typical efficiencies of up to 14% at 0.28 mm thickness.

Figure 2.4 Multi-crystalline silicon solar cell after screen printing.

2.5.2.1 Key issues and challenges
The advantages of crystalline silicon technology are as follows:

- There is a well-established and market-dominant silicon industry.
- It is possible to share research results with the microelectronics industry.
- Silicon is widely available for nearly unlimited solar electricity conversion, albeit in the form of SiO_2.
- Material quality is uniform (electrically, mechanically and chemically).
- The technology is environmentally benign and non-toxic for the whole production chain up to the solar cell.
- It is proven technology, and no complex processing is involved.
- Production can be easily automated or upgraded thanks to the modular nature of the production equipment.
- Conversion efficiency is relatively high compared with current thin-film technology.
- Cells have very good stability when they are properly encapsulated; the lifetime can reach 30 years.

Issues today are:

- The electronic grade (EG) silicon feedstock (shortage threat), ingot growth and slicing processes are expensive.
- The substrates are relatively thick, and have to be cut from an ingot of an expensive material.
- Cells are restricted to specific sizes, and have to be tabbed and interconnected. This makes monolithic integration into modules impossible. Automation is not yet sufficiently implemented.

2.5.2.2 Availability of cheap silicon at solar grade quality
Even though crystalline solar cell technology has achieved a high degree of maturity, further improvements can be expected, in both device and manufacturing technology. However, at today's growth rate and volume of production, problems with the availability of cheap feedstock materials will occur very soon. In this respect the problem is not so much the availability of feedstock (about 5% of the world's silicon production is used by solar cell production), but rather the ability to purchase it at a price that is low enough. The price for off-specification material, unused by the electronics industry, has increased to approximately €25/kg. At an average yield of 10–14 kg/kW$_p$, the feedstock contributes €0.25/W$_p$ to the cell costs.

There are the following options for solving the materials shortage problem:

- **Improved production methods for thinner wafers and improved solar cell efficiency** to achieve better material yield. The target should be about 7 kg/kW$_p$ at unit costs that should not exceed €0.20/W$_p$.

- **Ribbon or edge defined film-fed growth or foil casting of silicon melt.** This may lead to a doubling of the area of wafers produced from the same amount of feedstock at half the cost per wafer, as sawing is no longer required.
- **Investment in processing technology for solar-grade silicon feedstock.** Such a feedstock factory would commercially be viable only at capacities of about 1000–2000 t/yr. The necessary investment costs would require a consortium of PV manufacturers to agree on guaranteed demand for wafers or feedstock.
- **Significantly higher yearly growth rates for production of thin-film technologies.** If PV production continues to grow at a rate of 20% per year, and the supply of silicon feedstock is limited to a contribution of about 500 MW_p, thin-film production volume must grow at a yearly rate of 45% for several years to come.
- **Development of concentrator systems.** Such systems make better use of the expensive cell by concentrating sunlight onto it. This could be the lowest-cost option, especially for PV plants >10 kW in regions with high direct radiation. However, the technology for reliable low- and high-concentration systems has still to be developed.

2.5.2.3 Feedstock material

Today, most silicon cells are made from reclaimed material from the microelectronic industry. This situation may continue for a few years because of the expanding electronics industry, and also because the wafers used for PV are getting thinner and the efficiencies are increasing. Advanced slicing results in a pitch (the sum of wafer thickness and kerf loss) below 400 μm, and conversion efficiency above 17% is realistic. The consumption of silicon can therefore be reduced to less than 10 kg/kW_p, corresponding to a silicon ingot cost of €0.2/W_p (based on a cost of €20/kg). However, despite these efforts to optimize silicon usage, it is clear that a dedicated solar-grade silicon production facility is urgently needed to secure the supply of silicon for the PV industry.

Four main routes for the production of solar-grade silicon are currently being investigated (see Figure 2.5):

1   purification of metallurgical-grade silicon
2   direct carbothermic reduction of silica
3   use of other silicon waste streams
4   variations on the current poly-Si technology that are optimized for solar-grade silicon.

These routes differ greatly in terms of risk and potential. Route 4 has a relatively high chance of technological success, but a production plant would require very large investments, and the potential for producing solar-grade Si at low cost is relatively modest. Routes 1 and 2 carry larger risks for technological success but, if successful, might lead to a low-cost production process.

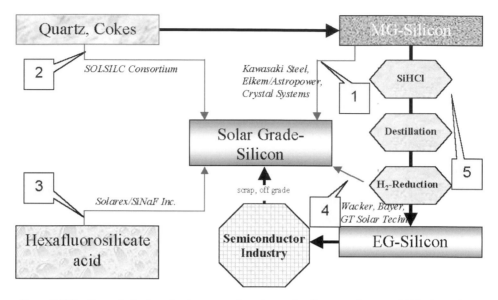

Figure 2.5 The four principal production routes for solar-grade silicon: purification of MG silicon, direct carbothermic reduction of silica, the use of other waste streams, and variations on the conventional processes (MG: metallurgical grade; EG: electronic grade). The thick arrow (route 5) shows the current process technology.

INGOT FORMATION AND WAFER PRODUCTION

Since both mono-crystalline and multi-crystalline silicon cells can currently be produced at about the same cost per W_p, ingot formation techniques for both processes should be developed further. In the long term, multi-crystalline casting processes show more promise in terms of throughput; but mono-crystalline cells are slightly more efficient and can be made thinner.

Mono- and tri-crystalline Si ingots are currently prepared by the Czochralski process. (Tri-crystalline ingots are Czochralski-grown crystals with three crystal orientations in longitudinal directions). The required developments concern the reduction of consumables (crucibles, gases, etc.) and the increase of throughput and yield.

Multi-crystalline silicon fabrication techniques can be subdivided into two groups:

- **Directional solidification in the melting crucible and block-casting.** This technique has the best potential for higher throughput and possibilities for scaling up. Production speed could be increased by using larger ingots or continuous casting methods.
- **Electromagnetic casting.** The advantages of this technique are the absence of crucible wear-out and melt contamination by the crucible, and the potential for higher throughput. Large-scale production of silicon using this method should be demonstrated, and the homogeneity of the material should be improved. Also, the material has smaller grains and therefore more grain boundaries to be passivated.

SLICING

Multi-wire saw cutting is the standard technique. The advantages over inner diameter saws are the possibility for slicing thinner wafers with low kerf loss, little surface damage and low breakage at relatively low cost.

The evolution towards thinner wafers imposes the requirement of appropriate handling and of manufacturing equipment that allows processing with high mechanical yield. A pitch (thickness and kerf loss) of less than 300 mm is a realistic long-term goal.

RIBBONS/SHEETS

The manufacturing of Si crystals by pulling ribbons out of the melt eliminates the losses of cutting and grinding. Efforts are needed to increase the material quality, throughput and solar cell efficiencies. The challenge is to produce and handle 100 μm thin ribbons and the solar cells made from them.

2.5.2.4 Cell processing

The cost goal at this level is about €0.35/W_p. This necessitates cell processes with low material cost and with a minimum number of operations (<10), whereby each operation takes less than 1 second. Moreover, these requirements should be met without compromising the cell efficiency.

The required future efficiency goals for industrial cells are 18–20% for mono-Si and 16–18% for multi-Si. The full squareness of multi-Si, which can achieve a higher packing density within a module compared with the semi-square mono-Si wafers, compensates for the lower efficiency. Note also that the efficiency of laboratory cells is still 3–5 percentage points higher than that of industrial cells. This difference is mainly due to:

- material quality and cell dimensions
- the need to use production methods with high throughput and low costs
- difficulty in implementing efficient texturing schemes
- achievable line widths of metallization.

This important difference illustrates the large potential for improvement in industrial production, which can be best obtained through process optimization using pilot lines where process parameters can be flexibly varied.

The developments needed in order to reach the goals for efficiency and cost in industrial production are summarized below:

- **Substrates.** Larger (10×20 cm^2, 15×15 cm^2, or even 20×20 cm^2) and thinner wafers (100–200 mm and below) with specially developed production processes. The challenge is to maintain a low series resistance while increasing the cell area, and maintaining high overall yield.

- **Optical confinement.** Techniques used are chemical texturing, laser texturing, mechanical texturing and photo-electrochemical texturing. The challenge is to obtain good texturing in an economical way, especially on multi-crystalline and thin substrates.
- **Junction formation.** Novel cell processes, such as the buried contact cell or advanced screen-printed cells. Further work is needed to increase the throughput of emitter formation and to integrate a selective emitter in a cost-competitive way.
- **Front contact formation.** Laser or mechanical scribing of a pre-diffused and nitride- or oxide-covered wafer, followed by a self-aligned electroless deposition of Ni, followed by Cu. Advantages are the fine line width and the large height/width ratio that can be achieved.
- **Further improvement of screen printing technology.** The use of novel screens and pastes, infrared firing and tighter control of the production environment could help to achieve line widths down to 50 mm, giving a reproducible fill factor approaching 80%.
- **Alternative printing techniques**, such as gravure offset printing, have been proposed but not yet demonstrated for solar cells. Offset printing could considerably improve the aspect ratio of conventional screen printing for small finger width.
- **Processing with low thermal budget.** Homogeneous emitters without etch-back can already be delivered by rapid thermal processing simultaneously with a back surface field and/or with surface passivation. Further research will lead to the processing of complete, high-efficiency solar cells in a single, short thermal cycle.
- **Use of manufacturing science in industrial cell processing.** Statistical process control and design of experiments are tools for achieving a cost-effective and robust process with minimal sensitivity to unavoidable small changes in the production environment. This applies in particular to computerized in-line measurement and control.
- **Environmental aspects of processing.** More attention should be paid to recycling chemicals and reducing the production of chemical waste production by using unconventional wet and dry chemical processes. Recycling complete modules and cells after their useful lifetime needs further study, in line with future environmental regulations.
- **Crystalline silicon cells with different designs** to increase cell efficiency, to simplify cell interconnection, to allow the use of large wafers, and to increase the visual appeal of the modules.

2.5.2.5 Cell production technology

A major issue for all solar cell technologies is now the development of manufacturing processes that achieve high volume, high throughput, high yield and high efficiency at low cost. The size of the current crystalline silicon solar cell factories with annual production volumes of up to 50 MWp allows some higher capital investment to achieve these goals.

The current cost of conventional solar cells, and in particular thin-film cells, is highly dependent on the capital cost of the necessary equipment. High priority should be given to developing integrated, low-cost processes from raw material to a solar cell (as has been tried by the Spheral[1] Cell). This development can be made together with other industries, such as the magnetic media or packing industry.

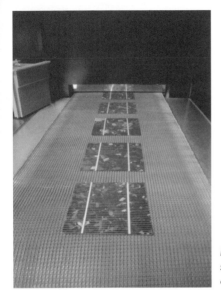

Figure 2.6 Production of multi-crystalline solar cells: after the screen printing of the contact grid a firing step is performed.

2.5.3 Thin-film crystalline silicon

Thin-film technologies have significant potential for cost reduction. To realize this potential, R&D efforts must be devoted to increasing efficiency and increasing acceptability through a certification procedure, recognizing the ability of thin-film technology to render services equivalent to, and cheaper than, those of crystalline silicon technologies.

Thin-film technologies not only reduce the quantity of silicon used but also allow more relaxed silicon specifications for a given efficiency. Also, after the wafer has been produced, subsequent processing can be very similar to that of conventional, crystalline silicon wafers. A major challenge is still to allow for sufficient absorption in the relatively thin films: this can be achieved by optical confinement together with texturing of the front and rear surface.

The principal approaches in producing thin silicon devices are as follows:

- **Deposition of Si on a ceramic or other low-cost substrate.** Efficiencies of 12% have been achieved. A particular case is the deposition of silicon by plasma-CVD on glass substrates, where 10% efficiency in the laboratory has been achieved on small-area cells.
- **Deposition on a relatively expensive substrate.** The substrate is reused, which requires that the grown silicon film be removed. For this purpose interface layers

between the substrate and the silicon film are used, to allow the film to be removed (lift-off) by either mechanical or chemical means. Efficiencies of more than 14% have been demonstrated.

Figure 2.7 Flexible solar cell: crystalline silicon thin film (few tens of microns) obtained by lift-off from a carrier substrate (IMEC).

Ultrathin (20–50 μm) self-supported crystalline silicon wafers or ribbons have the attractive feature of an increased mechanical flexibility (compared with a 100–150 μm thick Si wafer) and are therefore amenable to solar cell preparation by film transfer. Cost-effective and reproducible film generation, separation, transfer and cell manufacturing technologies have still to be developed.

For crystalline silicon thin-film solar cells on low-cost substrates the main challenges and problems to resolve are:

- reaching a sufficient crystal quality without an extra recrystallization step, at a high deposition rate
- finding cheap substrates that do not lead to excessive contamination of the deposited Si film by out-diffusion of impurities
- assessing the need for conductive substrates in relation to the module technology
- realizing an efficient bulk passivation technique that results in diffusion lengths higher than the layer thickness
- achieving a low surface recombination velocity at the Si layer/substrate interface
- realizing effective optical confinement schemes to maintain high short-circuit currents and to reduce film thickness
- in the case of low-temperature, plasma CVD deposition, improving throughput and material quality
- improving the deposition of high-purity silicon on low-grade silicon wafers, including MG silicon, for example by introducing barrier layers.

More research should also be focused on techniques to reuse the substrate, which can be of high quality. A major challenge is to remove the substrate. More experience has still to be gained on lift-off techniques for the substrate, both for chemical and for mechanical methods.

2.5.4 Amorphous silicon

Among thin-film solar cells, those based on amorphous silicon and related alloys (a-SiGe, a-SiC) are the most advanced. The technology for basic single-junction and tandem cells is mature and fully commercialized. For indoor low-power consumer products such as pocket calculators and watches, a-Si cells are used almost exclusively. a-Si modules are increasingly penetrating the market for consumer power (<50 W) such as small garden lanterns, battery chargers, and electric fences. There are about 20 production facilities in Asia, the USA and Europe, with capacities around 0.5–3 MW. Larger plants are under construction. Theoretical conversion efficiencies are around 18%; demonstrated small-area (<1 cm^2) efficiencies have been around 13% for single-junction and close to 14% for triple-junction stack structures.[2]

No major improvements in recorded initial efficiencies have been achieved for several years, which suggests that the practical efficiency limit for a-Si based solar cells is around 14–15%. The potential for scale-up has been demonstrated, with initial module efficiencies around 12%, relatively close to small-area efficiencies. Improvements in reducing performance degradation have led to stabilized module (30 × 30 cm^2) efficiencies of 10.2%. Commercially available modules offer stabilized efficiencies of 5–6%.

Since 1995, improvements have been made on triple-junction large-area modules, with lengths of up to 4 m available. These modules appear to exhibit stable efficiencies of 9–10%.

KEY ISSUES AND CHALLENGES
The advantages of amorphous-silicon-based solar cells are as follows:

- Material consumption is low, because a-Si has a high optical absorption coefficient and therefore cells can be made very thin (<1 μm), and only relatively small quantities of non-toxic raw materials (Si, C, Ge) are needed.
- Energy consumption for module production is low, due mainly to low-temperature processing.
- Module-related energy payback times[3] well below one year may be achieved with advanced large-scale integrated manufacturing processes.
- Cells and modules may be produced in arbitrary shapes, sizes and designs for integration in facades, rooftops or tiles. They may be designed opaque or semi-transparent. See Figure 2.8.
- Large-area thin-film technology has the potential to reduce cost dramatically. Projected module costs for large-scale production (>50 MW/yr) are €1/W$_p$ or lower.

Issues today are as follows:

- Vacuum processing is used, either in continuous lines or as a batch process. High-purity feedstock gases and cleanliness for the deposition process are required. Multi-layer and multi-junction structures for the more advanced cells need highly accurate

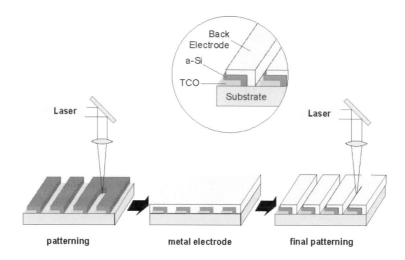

Laser   Laser

patterning   metal electrode   final patterning

Figure 2.8 Principle of patterning process in production of monolithic amorphous silicon. First transparent, conductive oxide (TCO) is deposited on the glass substrate, followed by the three active p-i-n layers. The final layer is the rear contact. Between each layers a patterning process is applied in sequence, such as single stripes (~10 mm wide) that are connected in series to each other.

and complex process controlling. Because of the complexity of the technology, the investment costs for equipment are rather high.

- In addition to extrinsic degradation mechanisms (such as layer interdiffusion and pinholes), amorphous Si and its alloys are inherently unstable. Exposure to sunlight produces recombination centres, with the result that the initial performance of a cell degrades by typically 10–25% until a stable state is reached, typically after a few months. However, this effect is reversible when the device is held for 100–300 h at a temperature of about 65°C. In operation, an amorphous module achieves an equilibrium state between these degradation and annealing process.

- The efficiency of the stabilized module is relatively low compared with that of crystalline solar cells. Commercially available modules have 5–6% efficiency; the highest module efficiencies achieved with advanced multi-layer structures are around 10%. The prospects for achieving more than 12% stabilized module efficiency are poor. It is a tough challenge to achieve 10% module efficiency in a cost-effective, high-yield production process.

RESEARCH AND DEVELOPMENT GOALS

Amorphous silicon thin films, which have the longest history of thin-film photovoltaics, have reached a mature stage. The low efficiency and initial degradation do not constitute a severe barrier to applications on buildings, such as roofs or facades. However, so far production costs have not been demonstrated to be significantly lower than those of other technologies. The solar cell efficiency is modest, and the production volume of current a-Si production plants is far below that necessary to justify vacuum systems of the size used for example in the glass-manufacturing industry, which can be operated cost-effectively. If complex vacuum systems can be operated cheaply enough, multi-layer amorphous substrates are the way to increase efficiency and stability.

Although intrinsic degradation is, in principle, unavoidable, the material stability *can* be improved slightly. Research should focus on improving deposition processes to reduce the effective gas consumption and increase the deposition speed.

Carrier collection can be enhanced and carrier recombination reduced by:

* reducing cell thickness, with correspondingly higher electric fields
* SiC buffer layers at the interface between the p-type and the intrinsic layer to prevent back-diffusion of electrons
* graded doping profiles across the i-layer to improve electric field distributions in the degraded (stabilized) material.

Optical absorption can be enhanced by light trapping with textured front and back contacts or textured substrates. Therefore research should be intensified for textured TCO layers such as SnO_2:F, or alternatives such as ZnO.

The equipment market should be carefully observed. Deposition equipment used for the first generations of thin-film displays may soon become available at considerably lower cost than that of custom-made equipment. This would allow further technology development of multi-junction modules at low capital cost, even though not yet at the largest size.

2.5.5 Thin-film polycrystalline technologies

The situation is different in the thin-film polycrystalline technologies, using $CuInSe_2$ or CdTe. Both materials can be produced at high efficiency and good stability. However, with the exception of sputtering, equipment has to be developed specifically for this purpose. Also, as with other thin-film technologies, problems of homogeneity, contacting and reliability have still to be resolved. It will be necessary to understand in more detail the losses in efficiency that occur when larger areas are deposited, and to develop remedies against them.

2.5.5.1 Cu(Ga,In)(S,Se)$_2$ (copper-indium-diselenide)

STATUS AND PROGRESS

During the last decade of the 20th century efficiencies of thin-film solar cells based on multinary chalcopyrite compounds of the type $Cu(Ga, In)(S, Se)_2$ improved continuously, and they have achieved the best results of all thin-film devices – close to 19% on a laboratory scale with cell areas below 4 cm^2. The best results have been achieved by depositing the chalcopyrite film using thermal co-evaporation of the elements at substrate temperatures around 550°C. Module efficiencies have already reached 12–13% for aperture areas of 0.1–0.4 m^2. Stability tests so far do not show any degradation of device performance.

KEY ISSUES AND CHALLENGES
The principal advantages are:

- the high degree of freedom for optimal adaptation to the solar spectrum due to easy tuning of electronic parameters by material variations
- the high conversion efficiencies – above 19%
- the absence of degradation in performance during lifetime
- the low material costs, as films can be very thin (less than a few microns)
- the low toxicity of both the device materials and the final product
- the high scalability of processes for large-area fabrication with high throughput
- the low energy consumption of fabrication processes and therefore the low energy payback time
- a module design that is adaptable to consumer application thanks to flexibility of patterning for monolithic integration of the module
- the high radiation resistance under space conditions.

The issues today are as follows:

- a technology for film preparation that is rather complex
- a scalability to high capacities with high yield that is not yet proven
- the availability of raw materials.

RESEARCH AND DEVELOPMENT GOALS
Fundamental R&D has to be continued intensively, and the results have to be transferred for scaling up. The main points are as follows:

- material research and modification (e.g. related chalcopyrites and other polycrystalline semiconductors, Cd-free films)
- modifications of device structures (e.g. superstrate)
- process simplification and patterning-related aspects
- development of strategies for recycling modules
- dry junction formation without CdS
- realization of cells with high voltage and low currents
- development of cells on flexible and lightweight substrates for special terrestrial and space applications.

The goal over the next years is to develop process conditions and deposition techniques with high yield to realize module efficiencies of more than 12% on areas around 0.5 m^2. The key points are absorber deposition and herein substrate temperature, simplification of laboratory techniques, and the achievement of high tolerances of process conditions with respect to device performance, material yield and throughput. Furthermore, there is a need for developing in-line processes and quality control (e.g.

criteria for substrate/back contact quality, performance of absorber layer). All different techniques have to prove their suitability for further scaling up to areas of 0.4–0.8 m² with pilot plant capacities in the low MW/yr range. Environmental, health, safety and energetic aspects during raw material processing, module production and throughout the lifetime of the product have to be taken into consideration when large volume manufacturing is to be realized. It will be important to develop strategies for the recycling of these modules at the end of their useful life (30 years).

2.5.5.2 *CdTe/CdS*

STATUS AND PROGRESS

Intense research activities have led to substantial progress in the development of highly efficient CdTe/CdS cells. Different deposition methods with promising economy have been applied. Besides closed space sublimation and chemical deposition methods, spraying, electro-deposition and sintering of screen-printed films have been applied successfully. Laboratory cell efficiencies of 16% have been reported that prove the potential for economy, especially in view of the simple basic structure.

Existing production is for consumer applications only. Energy payback time is assessed to be in the order of 1–2 years. CdTe/CdS modules operating under both external and internal accelerated tests show stable efficiencies. Modules manufactured by industry now reach efficiencies of more than 10% on 0.5 m² sizes.

KEY ISSUES AND CHALLENGES

The general advantages of CdTe are as follows:

• ease of growth
• different deposition methods with promising economy possible
• almost optimal bandgap (1.4 eV), allowing very thin structures.

Issues of CdTe and CdS are as follows:

• toxicity of Cd, Te and their chemical compounds used for preparation: toxic compounds could potentially contaminate the environment during fabrication, operation or recycling.
• limited availability of compounds used: the supply of Te has to be substantially increased if CdTe PV modules are to cover even a small amount of the current global consumption of electricity.

RESEARCH AND DEVELOPMENT GOALS

Although polycrystalline CdTe solar cells have demonstrated a high efficiency, a major difficulty has been in obtaining sufficiently high p-type doping to lower the bulk and contact resistances. Basic R&D effort is needed to improve cell characteristics and increase cell efficiencies, and in particular further improvement of the p-type conduction of CdTe, leading to more stable *I–V* characteristics. This can be improved by an annealing process in oxygen.

The most important objective is to increase yields in large-scale module production, by improving and simplifying the manufacturing processes and decreasing the production costs. The most important fabrication criteria are as follows:

- high material utilization rate
- low-cost equipment
- low energy requirements
- high specific throughput
- high growth rates
- good process control
- wide growth window
- low processing temperature
- low vacuum.

Significant progress has been made using the rapid, large-area closed space sublimation (CSS) technique. Some criteria, such as non-expensive equipment, low energy requirements and simple manufacturing techniques, can fulfilled by sintering techniques and spray pyrolysis, which still offer considerable potential for technical improvement.

The environmental aspects of photovoltaic module manufacturing, especially in relation to CdTe and CdS, demand a minimum of material use, high yields and safe handling, as well as recycling of wastes. It is believed, however, that reduced material quantities in the deposition processes could overcome this problem. Concepts have to be developed for disposal and recycling of modules at the end of their lifetime.

2.5.6 III-V cells

STATUS AND PROGRESS

Cells made from compounds of elements in group III and group V of the periodic table, such as gallium arsenide, have achieved the highest laboratory solar cell efficiencies reported to date, both for non-concentrator cells (approximately 28%) and for concentrator cells (200 suns, about 32%). They have reached practical efficiencies close to the physical limits for both one-sun and concentrator illumination. Improvements in efficiency are still possible as a result of design refinements and of improvements in material and processing technologies.

Gallium arsenide (GaAs) can be considered to have an advanced laboratory status and to be progressing towards maturity at the industrial stage: several companies in the USA, Japan and Europe are able to produce these cells, which are used almost exclusively for space applications. But as GaAs efficiencies are not significantly higher than those of silicon cells manufactured with comparable effort, it is unlikely that this material can compete in cost with concentrator Si cells in the near (<10 years) future.

However, the potential of III-V semiconductor solar cells is shown to full advantage when very high-efficiency solar cells are required, alone or in combination with concentration, together with a more efficient use of the solar spectrum. In this way, the insight gained into the working principles of solar cells at high current densities and matched to different parts of the solar spectrum is of great importance in developing all solar cells for terrestrial applications.

KEY ISSUES AND CHALLENGES

The advantages of III-V cells are as follows:

- Very high efficiencies are possible.
- It is possible to develop multi-junction devices.
- They are suitable for high-concentration application (>500 suns).
- They have low efficiency loss with increasing temperature.
- Their degradation under particle radiation conditions in space is low.
- There are spin-off applications in optoelectronics.

Issues of today are as follows:

- the high cost of substrates
- the use of dangerous components
- the need for operation under concentrated, direct sunlight for optimum performance, which limits the application potential.

RESEARCH AND DEVELOPMENT GOALS

The goal for III-V compounds is to achieve super-high efficiencies of >40%. This should be made possible by fabricating four-junction devices. These cells will be realized either by mechanically stacking single or monolithic tandem devices or by total monolithic integration. In any case, new materials (ternary or quaternary) have to be developed and characterized with respect to their suitability as photovoltaic absorber materials. The big challenge is to fabricate high-quality material while designing the band-gap. Lattice-matched and lattice-mismatched materials have to be investigated. For the application of monolithic tandem cells in high-concentration systems, the development of high peak current, low resistance and low optical absorption tunnel diodes has still to be improved. For monolithic structures the challenge is to grow in the same structure two devices of different materials with both lattice and current match as well as growing a suitable lattice-matched tunnel diode for interconnection of the two devices.

Another challenge for the III-V compounds is to grow the structure on a reusable substrate in order to reduce cost. Special lift-off techniques have to be developed.

2.6 ADVANCED RESEARCH AND CROSS-FERTILIZATION WITH OTHER TECHNOLOGY DEVELOPMENT

Progress in technology is realized largely through incremental improvements of available technologies. This is true for photovoltaics as well. The incremental improvement has to be cost-effective: that is, the 'cost of ownership' of the manufacturing equipment that implements this improvement has to be at least offset by gains in the price/cost relation. Much of the R&D aiming at these technology improvements will be done by industry itself in order to gain competitive advantage.

In contrast, advanced research aims to explore new approaches to photovoltaic devices that go beyond currently available technology. Cost reduction and efficiency improvement might well be possible with inventions that are still to come ('surprise' technologies). Industry will take an interested look at these developments, but will not be able to finance such long-term, strategic research, as the time-to-market would be too long for private investment. Clearly there is a role for publicly funded research in this context.

The photovoltaic technology that is available today has benefited from the knowledge, equipment and materials of the semiconductor industry. This is also true for the development of thin-film amorphous silicon, which uses process equipment similar to that of today's display industry. However, it is far from the technical optimum that is theoretically possible. Silicon was and is the semiconductor of choice for solar cells, and is in its basic configuration limited to 26–27% efficiency.

We can therefore set out the major objectives for advanced research: novel solar cell materials have to be developed that are efficient, inert, low-cost, non-toxic and abundant; and fabrication processes are required that operate at moderate temperatures, do not need high vacuum, and are suitable for large-area manufacturing at high throughput and yield.

As for crystalline silicon technologies in the past 30 years of photovoltaics, new devices can speed up their growth into maturity by taking on board the basic understanding, materials, technologies and processes of other related developments. For thin films, current display technology will be the industrial driver, as well as high-volume deposition methods such as those found in the production of magnetic media. There is also a need for increased cross-disciplinary research among physicists and (electro-) chemists, to understand and exploit the effects of photons on different materials, and to explore the possibility of extracting electricity from these materials. Thin-film polycrystalline processing has still to exploit other, non-photovoltaic technologies.

This advanced research would ideally include research into the effects of storing energy in the same device as that it in which it has been generated by photons.

Ways to formalize cross-fertilization should be exploited. Financing could be adjusted to the amount of cross-over technologies involved. An example of such a cross-fertilization in another industrial area is the potential replacement of light-emitting diodes by organic semiconductors.

2.6.1 New research approaches for highly efficient photovoltaic conversion

Based on the extensive knowledge accumulated in recent years a screening study for new materials for high-efficiency thin-film cells should be made. It is time to start specific, basic research on the **ultra-efficient**, **low-cost solar cell**. The first aim should be to understand all phenomena related to photo-conversion in a variety of materials, in order to define 'virtual' materials that have the required properties. In this context, the modelling of intermediate-band solar cells capable of absorbing sub-band-gap photons without sacrificing cell voltage could be important.

2.6.2 Organic and polymer cells

Great efforts have been put into the development of cheaper thin-film solar cells, both those that consist of purely inorganic materials and those that contain organic materials. Examples of the latter are dye-sensitized photoelectrochemical solar cells, molecular organic solar cells (MOSC) made from relatively small organic molecules, and polymer organic solar cells based mainly on electrically conductive polymers.

These three types have an enormous potential for future photovoltaic applications for several reasons. The amounts and necessary purity of organic materials are small, and their large-scale production is considered to be relatively easy compared with that of most inorganic materials, offering low-temperature processing, cheap materials and flexibility. Thanks to the infinite variability of organic compounds, they can in principle be tailored to any needs, which makes them widely applicable.

The interest in **dye-sensitized solar cells** has increased enormously since their first realization in 1991. This photoelectrochemical cell is based on charge transfer from light-excited dye molecules to an inorganic semiconductor with a large band-gap. By using nanoporous TiO_2 with a large effective surface area, and attaching dye molecules to this area, enough light can be absorbed to achieve useful efficiencies. These devices have shown efficiencies up to 11% on small areas (0.25 cm^2) and 5–8% on somewhat larger areas (1–5 cm^2).

At this stage, fundamental and technologically oriented research are running in parallel. While many academic research groups are investigating the various unknown aspects of this solar cell, several companies and institutes around the world have concentrated their efforts on the technological development of efficient large-area multi-cell modules that are simple to make and stable over many years.

Molecular organic solar cells (MOSC), made in most cases from flat aromatic molecules, have been investigated since the early 1970s. Recently, a significant efficiency enhancement was demonstrated for a laboratory Schottky-type MOSC based on pentacene (efficiency up to 4.5%) after molecular doping with iodine.

The youngest and fastest-growing field in organic solar cell research is based on the use of **electrically conducting polymers** as photovoltaic materials. The inherent processing advantages of this technology, already developed for a number of thin-film technologies (LEDs and FETs), as well as the flexibility of chemical tailoring of desired properties, make solar cells based on polymers very attractive.

External power conversion efficiencies of up to 3% have been achieved for laboratory cells consisting of a bulk heterojunction of light-absorbing polymers such as phenylene-vinylene and fullerene (C60) molecules. Although several aspects remain to be investigated (such as conducting polymers with better light-absorbing properties, film-morphology, and device stability), these encouraging numbers appear promising for a cheap production process of efficient 'plastic' solar cells in the future.

2.6.3 Modules

2.6.3.1 Status and progress
About 35% of the module cost is devoted to module-related technologies. Therefore significant efforts are needed to contribute at this level to cost reduction.

Most of today's module manufacturers employ a standard laminate of the glass-EVA-tedlar type, and most experience has been gained on this composition. For integration into the built environment particular designs have been offered to the market, often as a double-glass structure, which allows partial transparency. With the standards for type approval well in place, module reliability is very high and meets the requirements of the markets.

2.6.3.2 Key issues and challenges
*   Processing costs for tabbing, stringing and lamination are still too high.
*   Some of the processes involved are difficult to automate.
*   Custom-made modules, as required in the building market, are complex.

2.6.3.3 Research and technology goals
Today's encapsulation processes, although very effective, are expensive and often limit the flexibility needed to produce market-tailored products. Development should be undertaken not only to improve current processes but also to look for alternatives that will be more suitable for a continuous production process. Large-quantity industrial products always produce large areas; the final product is cut from this large-area raw product. Today's crystalline PV module technology still has the opposite approach. Development of materials and cell technologies will be necessary to allow a fast process of module pre-assembly. A breakthrough would be possible if contacts for tabbing and stringing could be made on the same side of the wafer (using point contacts, wrap-around or wrap-through contacts).

Research will also be necessary into the options for electricity storage within the module compound. Current lithium-polymer technology, as known from cellular phones, seems to be a very suitable candidate for such integration. The integration of inverters or other electronics directly in the encapsulation systems is desirable and also possible, once all the required electronics can almost fit on one chip. This would include DC-DC converters that increase the voltage. Such modules would have considerably lower assembly costs.

Efforts should continue on standardization, reliability and environmental issues in order to make PV modules acceptable on the market as a mature product:

• In order to avoid disappointment for users, and to facilitate the financial planning of PV systems, a standard and realistic energy rating concept should be established.
• New encapsulants should be investigated, including resins that can be hardened by ultraviolet treatment.
• Life-cycle assessment of material and energy flows should further identify any environmental risks of PV modules. Also, the possibility of reprocessing recovered wafers from used or reject modules, aiming at important cost and energy savings, should be considered for all module technologies.

The further development of building-specific modules, combining solar module technology and roof-sealing techniques, should result in an architecturally, technically and economically satisfying product that meets both complex requirements and high standards of aesthetics. The testing and further optimization of existing concepts such as PV tiles and PV walls is necessary, together with the development of new approaches. The building industry and architects must be closely involved in this.

The potential for cost reduction by simplifying design, by using cheaper materials or by using mass production techniques has to be assessed.

The economy associated with the multifunctional integration and replacement of conventional building elements is an important consideration. Therefore, opportunities such as structural glazing, natural daylighting and combination with thermal applications should be taken up as much as possible. Often more than 60% of the costs of the photovoltaic installation can be recovered by the replacement of building components that need to be replaced anyway.

2.6.4 Concentrators

Photovoltaic concentrating systems will find their main applications in dry or island regions where the amount of direct irradiation is high. At similar costs to those of traditional flat-plate systems, concentrators have the advantage that a larger fraction of costs occurs in the non-PV technology components, and has the potential to be covered by local labour.

Research into concentrators is inherently linked to the following topics.

2.6.4.1 Structure

Two important requirements are low cost, including installation costs in the field, and reduction of deformations under the various stresses caused by wind and gravity to acceptable limits. Experience with a wide variety of structures is very limited, and it is necessary to make a sound comparative analysis. The accuracy of passive trackers, based on differential heating of fluids by the sun and piston movements, should be further improved.

2.6.4.2 Cells

Research into and development of concentrator cells must emphasize high efficiency, even if relatively complex processing is required. Another requirement for high-concentration solar cells is that the series resistance must be considerably reduced.

2.6.4.3 Bonding and connecting

The bonding of concentration cells must be able to extract large currents, which implies the need for low series resistance, an appropriate thermal expansion coefficient in relation to the semiconductor, and good thermal contact with a heat sink while maintaining good electrical insulation.

2.6.4.4 Cooling

Cell cooling is also a crucial aspect of concentrating devices. Both active and passive cooling have been used for this task. Passive cooling is by far the most cost-effective, and in most cases it is sufficient.

2.6.4.5 Optics

Further research is needed on geometric designs under conditions of mirror misalignment or deformation. Also, low-cost and reliable coating of the optical systems (when reflective) will lead to overall energy production improvement at lower cost.

2.6.4.6 Housing

The housing is the almost sealed box, containing the lenses and the cells, which is mainly for protection against dust and humidity. Further understanding is necessary for a reliable design at high irradiance and high currents in a humid environment.

2.7 SYSTEMS INTEGRATION AND TECHNOLOGIES

2.7.1 Technology status of components

2.7.1.1 Charge regulators

Low-cost devices (€30) now include deep discharge protection, temperature compensation and charge equalization cycles. Higher performance regulators include indication of state-of-charge and some basic monitoring function. There are devices on the market that allow paying schemes by smart cards.

2.7.1.2 Batteries

The valve-regulated lead–acid (VLRA) storage system is used in nearly all stand-alone application. Photovoltaic applications make up about 2% of the accumulator market. Accumulators are often optimized for photovoltaic applications, in particular in relation to the stratification effects of low charge and discharge currents. Even though sealed (gelled electrolyte) lead–acid accumulators often have a better life-cycle economy than

VLRA accumulators, the higher investment costs mean that their use is restricted to high-value applications such as telecommunication and navigation.

2.7.1.2 Inverters

The large-scale inverter market was stimulated by the high-volume building systems. The development of stand-alone inverters has led to higher reliability and slightly lower prices, but grid-connected inverters in the 1–3 kW range achieved a breakthrough in terms of costs, reliability, efficiency, safety and built-in intelligence. Throughout the technology, transistorized power stages are used. Larger inverters (>50 kW) did not improve figures-of-merit to the same extent, probably because most of them are custom-built and cannot profit from high-volume production.

Apart from module-integrated inverters, which are still to become cost-effective and fully reliable, string inverters are appearing on the market. Both concepts aim to achieve further progress as a result of the high production volumes that the modular concepts are expected to allow.

2.7.2 Generic R&D needs for systems

Further research is required in three areas:

- lowest-cost integration in the field, with a high priority given to integration into buildings (but there are many other market segments that would profit from new developments in system integration, in particular in applications for rural or underdeveloped areas)
- electricity storage, with a focus on making new technologies (Li-polymer, flywheels, supercapacitors, hydrogen fuel cells) suitable for PV use
- improvement of system components, such as charge regulators, maximum power point trackers (MPPT), DC/DC converters and DC/AC inverters, and at the level of intelligent and extremely low-cost control and supervisory systems. In particular, for hybrid wind–diesel–PV systems, where markets require a high degree of adaptability, self-learning systems would need to be further developed. Especially important for further developments is the design of application-specific integrated circuits (ASICs) aiming to reduce the costs of balance-of-system components by miniaturization.

2.7.3 Goals for technology development: systems integration

2.7.3.1 Grid-connected systems

Integration of PV systems addresses the properties of PV systems technology in view of the diversity of local contexts, mainly integration within buildings and utility grids in Europe but also avoidance of grid extensions. Each additional kilometre of overhead gridline can typically consume half a hectare of forest.

The overall task of decentralized supply is to be addressed through all its phases, from development to installation, expansion, operation, maintenance, repair and recycling. The following promising approaches are at the beginning of their development:

• supervisory control methods for flexible integration
• on-line concentration of distributed monitoring data
• grouped installation, expansion and compounding of plant complexes.

Especially in view of eased financing and secured operation, further work in these fields is a priority.

Utility interconnection issues have been widely addressed, and could to a large extent be solved satisfactorily in line with existing standards and norms. For integration into the built environment, in addition to adapted PV modules and mounting systems, the main topics are:

• the combination of thermal and photovoltaic use
• an aesthetically pleasing appearance, and acceptance by the public
• design rules for architects
• simplified mechanical and electrical installation.

Even though considerable progress has been made, integration with rational use of energy, tariff structures and supervisory control need more effort. Acceptance will be further improved by providing clear information on yield, and by timely repair of malfunctioning solar systems.

Figure 2.9 This laboratory building at the European Commission's Joint Research Centre in Italy has been retrofitted with amorphous Si modules. With an active area of 540 m² it is rated at 21 kW$_p$. After five years of continuous operation, efficiency has stabilized at 4%. (JRC)

2.7.3.2 Stand-alone systems

Photovoltaic off-grid electrification in developing countries is a distinct issue. It differs from integration in industrialized countries, and often demands the establishment of an appropriate infrastructure. R&D is required at the system level to meet the criteria of:

- modularity and extendibility of stand-alone systems ranging from about 10 W_p
- the use of components that are locally available, and cheaper than imported components
- high reliability in tropical, semi-arid and arid climates, robustness and shock resistance
- low price and simple operation, allowing for local production or repair.

PV village-grid systems to date have been successful in only a few cases, and reliability is still crucial before these systems can be implemented on a large scale. The focus for the future should be on the relationship between PV technology and the actors involved in related applications, as well as on the corresponding organizational and financial issues:

- manuals and tools for training of users and skilled staff, adapted to local culture
- PV installation, maintenance and repair schemes for local infrastructures
- financing schemes adapted to the local microeconomic structure.

Technical innovations, e.g. counters and supply switches for prepaid cards or electronic money, are coming mainly from synergies with information and communication technologies in other areas.

2.7.3.3 Supply guarantee

For commercial applications such as telecommunications, and most residential and public applications, the acceptance of PV energy can be achieved by guaranteeing reliability of supply.

R&D progress has progressed in:

- accreditation and quality standards
- system layout procedures
- cheap and reliable monitoring techniques for on-line system error detection
- maintenance and repair schemes.

However, the two last topics in particular are not sufficiently satisfied and need further efforts towards a reasonable standard that goes beyond one-off solutions.

Quantitative interpretation of monitoring results is still missing from most systems, and there is still considerable untapped potential for system optimization. A new approach together with the previously mentioned issue of supply guarantees could be promising.

2.7.4 Goals for technology development: electricity storage

With respect to stand-alone systems, R&D needs to focus on electricity storage and combination with other renewables. Families of similar small systems for different purposes but with as many common properties as possible allow the achievement of the scale effect, but some development is still needed, in particular in relation to quality standards. For electricity storage, batteries often do not live up to the expectations of either the designer or the user. Improvements need to be made in order to establish adequate design specifications and procedures as well as the active (built-in) limitation of charge and discharge.

Considerable effort is still needed to adapt new accumulator developments such as NiMH and Li-ion gradually to the PV market, as far as costs allow. Also, very recent results achieved with supercapacitors, flywheels and pump storage must be backed up by further research efforts.

2.7.5 Goals for technology development: components

By fully integrating a grid-connected inverter into a module, it becomes an independent grid-connected unit. This new approach has considerable cost reduction potential (up to 20%). R&D efforts should focus on:

* improving the design of an AC module inverter and developing a more cost-effective connection system suitable for large-scale applications in buildings
* testing the performance of AC modules, with special attention paid to thermal stress (due to the placement of the inverter at the rear of the module), current harmonics and possible interaction.

2.7.5.1 Modularity and standardization

By means of a modular systems technology using a limited set of flexible units, most diverse supply systems can be configured, e.g. for stand-alone applications. In addition, the use of contemporary microprocessor technology in combination with communication bus structures makes it easier to modularize supervisory control. The following are still needed:

* compatible fieldbus addressing of components from different manufacturers
* standardization of monitoring and supervisory control
* power electronics control intelligence for off-grid use.

2.7.5.2 Technical quality

For the balance of system components and the appropriate electrical appliances, the quality of components and systems and the repair and maintenance procedures have to be addressed by new standards in type approval testing and need further work, in particular on meaningful, effective test methodologies.

2.7.5.3 Power conditioning

The tendency towards self-commutated, parallel, low-power inverters, corresponding to the modular construction of photovoltaic electricity generation, has progressed well in the last decade. They are capable of reducing harmonics, and allow both parallel and stand-alone operation. The cost reduction, correlated to the accumulated production, can be of similar magnitude to the anticipated cost reduction of PV modules.

Major efforts in research should be directed towards:

- miniaturization and reduction of losses of power electronics
- inductive PV module coupling with DC transformers to overcome the drawbacks of feeding cables through the module encapsulation and to decouple mechanical and electrical installation
- supervisory control methods for operation in all sensible electrical system configurations
- the use of power-line-modulated communication techniques
- methods for robust pulse inverter parallel operation without external synchronization.

The development of application-specific integrated circuits requires a considerable amount of pre-industrial work. However, in order to find successful solutions that can be incorporated by many manufacturers, industry will have to be involved early in the developments.

ANNEX

All electrical performance data for photovoltaic modules are referred to standard test conditions (STC), defined by a temperature of 25°C, an irradiance of 1000 W/m^2, and a standardized solar spectral distribution of air mass 1.5. When the power of photovoltaic devices is measured at these (often artificial) conditions, the unit used is **peak watt** (W$_p$). The electrical characteristics of solar cells are best described in an **I–V** plot, in which current is plotted as function of the external voltage. There are three distinct points on such a curve (Figure 2.10):

I_{sc} **Current at short-circuit.** This is the maximum current that a solar cell can generate; the voltage across the cell is zero. The short-circuit current is a measure of the quantum efficiency, and indicates to what extent photons can generate electron–hole pairs. It is directly proportional to the illumination intensity.

V_{oc} **Voltage at open circuit.** This is the voltage that a solar cell exhibits when no load is connected; the current is zero. The open circuit voltage is correlated to the lifetime of the carriers during their diffusion through the material. The more carriers recombine before reaching the contact grid, the lower this voltage. The voltage is nearly independent of the illumination intensity; it follows a logarithmic dependence.

V_{mp}, I_{mp} **Voltage and current at maximum power.** Between the previously described extreme points one can designate that current–voltage pair of data where the delivered power reaches its maximum. This point also indicates the **squareness** of the *I–V* plot, which is given as the ratio of the inscribed rectangle of the maximum power to the external rectangle with the sides V_{oc} and I_{sc}. This ratio is the **fill factor**, an important figure of merit.

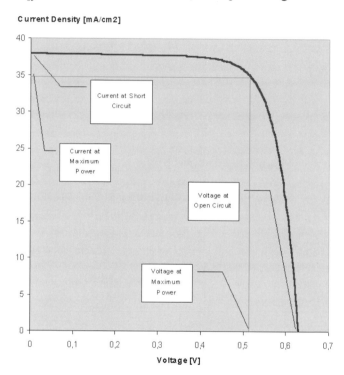

Figure 2.10 Typical current–voltage characteristics of a photovoltaic solar cell.

NOTES

1   Registered trademark of Texas Instruments.
2   To absorb a larger band of the solar spectrum, more than one solar cell can be stacked on each other. Any layer is transparent for wavelengths that are absorbed by the layer below. In the case of a-Si, however, tandem structures using layers with similar band-gaps are used to increase the stability of the cell.
3   Energy payback time is the time a photovoltaic module needs to generate the same amount of electricity as was consumed for the production of the module. The time for a-Si modules today is between one and two years.

3. SMALL HYDRO POWER

P Fraenkel (Marine Current Turbines Ltd)

3.1 INTRODUCTION

Hydro power is the largest and most mature application of renewable energy, with some 678,000 MW of installed capacity, producing over 22% of the world's electricity (2564 TWh/yr) in 1998.[1] In Western Europe, hydro power contributed 520 TWh of electricity in 1998, or about 19% of EU electricity (thereby avoiding the emission of some 70 million tonnes of CO_2 annually). Despite the large existing hydro power capacity, there is still much room for further development, as most assessments assume that only around 10% of the world's total viable hydro potential is currently exploited.

This chapter is limited to small-scale hydro power (SHP), since large-scale hydro power is technically mature and its R&D requirements – mainly refinements by the large specialist companies active in this field – are generally being well taken care of. Indeed, the development of SHP has tended to be neglected, partly due to the false but widely held impression, gained from the maturity of large hydro, that there is not much scope for technical development and improvement of any hydro power systems. There is, however, considerable scope for improving the cost-effectiveness of SHP, especially with low-head systems, through both technical and non-technical innovations.[2]

There is no general international consensus on the definition of SHP; the upper limit varies between 2.5 and 25 MW in different countries, but a value of 10 MW is becoming generally accepted, as, for instance, by ESHA (the European Small Hydro Association).[3] This chapter will therefore use the definition for SHP as any hydro systems rated at 10 MW or less. SHP can be further subdivided into mini hydro, usually defined as <500 kW, and micro hydro, <100 kW.

Small-scale hydro power (SHP) is mainly 'run of river': that is, it does not involve significant impounding of water, and therefore does not require the construction of large dams and reservoirs. Moreover, SHP mainly involves smaller manufacturers, generally small or medium-size enterprises. Few of the large companies that produce large-scale systems are very active in manufacturing and marketing small systems, even though some include small systems in their product range (usually simply scaled down from their larger systems).

Whichever definition is used, SHP is one of the most environmentally benign forms of energy generation, based as it is on the use of a non-polluting renewable resource, and requiring little interference with the surrounding environment. It also has the capacity to make a significant contribution to the replacement of fossil fuel since, unlike many other sources of renewable energy, it can generally produce some electricity at any time on demand (i.e. it needs no storage or back-up systems), at least at times of the year when an adequate flow of water is available. In many cases, it can do so at a cost competitive with that of fossil fuel power stations. For example, a 5 MW SHP plant typically displaces 1400 t/yr of fossil fuel, and avoids the emission of 16,000 tonnes of CO_2 and over 100 tonnes of SO_2 per year, while supplying the electricity needs of over 5000 families.[4]

3.2 POTENTIAL AND STRATEGIC SUMMARY

Small hydro power has a huge, as yet untapped, potential, which would allow it to make a significant contribution to future energy needs, and it depends largely on already proven and developed technology, although there is considerable scope for its further development and optimization. Table 3.1 shows the estimated distribution by region of global SHP capacity, together with two projections of possible future SHP development, for 2020, under alternative 'business as usual' and 'accelerated development' scenarios (adapted from the World Energy Conference committee study on Renewable Energy Resources).[5] The European Renewable Energy Study[6] envisages that hydro output for the 12 EU countries will increase from 165 TWh/yr to 185 TWh/yr (under the 'existing programmes' scenario) between 1990 and 2010; in the same period, the total electricity generated by renewable energy technologies will increase from 191 TWh/yr to 305 TWh/yr. This means that hydro's contribution to electricity generated by renewable energy technologies will shift from 86% to 60%, though by any measure this still makes it the largest contributor by far.

Table 3.1 Estimates of realizable small-hydro potential by region with two development scenarios (capacities given in MW)

Region	Baseline 1995	'Business as usual' 2020	Accelerated development 2020
North America	4861	6152	12,906
Latin America	1992	5751	6557
Western Europe	8822	12,587	21,692
E. Europe & the CIS	2801	3997	4197
Middle East / N. Africa	81	233	266
Sub-Saharan Africa	324	935	1065
Pacific	124	177	306
China	6963	20,101	22,915
Rest of Asia	614	1772	2021
Totals	26,582	51,705	71,925

In Europe, small hydro provides over 9 GW of capacity, which contributes over 30 TWh/yr to the EU's electricity supply.[7] There is an estimated extra 9 GW of European small hydro potential, including refurbishment projects that can reasonably be developed.[7] The European Commission has announced a target to increase small hydro capacity by 4200 MW (50% above the present level) by the year 2015.

Table 3.2 indicates one assessment of expected future growth in small hydro capacity, both within the EU and worldwide.[8]

Table 3.2 Growth in SHP capacity: 1980 to 2010

	1980	1985	1990	1995	2000	2005	2010
EU installed capacity (MW)	5900	6700	7700	9000	9600	10,300	11,000
World installed capacity (MW)	19,000	21,000	24,000	27,900	37,000	46,000	55,000

Recent growth rates for SHP have been disappointing compared with other forms of renewable energy development (less than 3% per annum improvement in capacity in Europe). This is because the development of SHP, both in Europe and elsewhere, has not received the same support that has been given to other forms of renewable energy. The following factors may account for this:

- There is a common perception that the technology is mature and fully developed, and that therefore it does not need any significant level of institutional encouragement or support (i.e. it is wrongly assumed that market forces alone will be sufficient to take it forward). For this reason, it is commonly excluded from programmes designed to assist other forms of renewable energy development. However, there is still, in reality, potential for development and improvement of SHP.
- Economic analyses of hydro power projects generally give no significant credit for the exceptionally long useful life and low running costs of SHP, and the high upfront costs tend to make it seem financially unattractive, unless low, discount rates are available.[9] The life of SHP systems might be 30–60 years, but the capital is commonly amortized over 10–15 years, making the electricity unacceptably expensive for that first period but inexpensive thereafter.
- There are numerous other institutional barriers, mainly resulting from the difficulties inherent in gaining permission to abstract water from rivers in many countries, and due also to perceptions that hydro plant might adversely affect fishing, boating and other leisure interests. In practice, though, well-designed hydro systems can avoid making any serious impact, environmental, social or otherwise. Difficulties in gaining affordable connections to the grid are also common, although this situation is tending to improve.

With a new international climate of concern about global environmental dangers, SHP clearly deserves to be more strongly promoted and more widely and effectively developed, yet progress so far seems disappointing to many who are involved in advancing its use.

3.2.1 Basic principles

The main requirement for hydro power is to create an artificial head of water so that water can be diverted through a pipe (penstock) into a turbine, from which it discharges – usually through a draft tube or diffuser – back into the river at a lower level. Various types of turbine exist to cope with different levels of head and flow. There are two broad categories of turbines:

- **impulse turbines** (notably the Pelton), in which a jet of water impinges on the runner, which is designed to reverse the direction of the jet and thereby extract momentum from the water
- **reaction turbines** (notably Francis and Kaplan), which run full of water, and in effect generate hydrodynamic 'lift' forces to propel the runner blades.

Figure 3.1 15 kW Micro hydro multi-jet Pelton installation in the Falkland Islands. (IT Power Ltd)

It is normal to achieve optimum energy conversion efficiencies with all types of hydro turbine in the range 70–90%. The larger the turbine, the higher its efficiency, generally speaking: hence good quality designs of several hundred kW or greater tend to approach or even exceed 90% optimum efficiency. In contrast, a micro hydro turbine of, say, 10 kW might have an efficiency of the order of 60–80%. Part-load efficiency varies considerably depending on turbine type; it is usually possible to achieve good efficiency over quite a broad power range, but to do so generally involves some cost and complexity.

Sizing of turbines depends on the flow characteristics of the river or stream to be used; generally, the maximum flow rate of a European river in flood can be several orders of magnitude greater than the minimum flow rate. The energy capture depends on the sizing strategy: the larger the turbine at a given site, the poorer its load factor (or capacity factor), as it will only run at rated power for a shorter period. A turbine sized to use the minimum flow only can have a load factor or capacity factor approaching 100%, but it will obviously extract less energy per annum than a larger turbine. A system optimized for cost-effectiveness needs to have a high load factor.

Resource assessment for hydro turbines has been developed to a high standard and should preferably be based on actual flow measurements, but can also be based on computer-based analysis of rainfall, catchment area and run-off.

3.2.2 Experiences to date

Small hydro has been exploited for centuries. Many tens of thousands of small-scale hydro power systems (i.e. watermills) were in regular use by the 18th century, right across Asia and Europe, mostly for milling grain. These ranged from simple Norse wheels (25,000 of which are still in use in Nepal) to sophisticated waterwheels, fitted with speed-governing mechanisms even in the 18th century, and often capable of as much as 70% efficiency.

In the 1880s, hydro turbines were first used to generate electricity for practical purposes, and in Europe the turbine took over from the waterwheel almost completely by the end of the century. Until the 1930s, small turbines were increasingly used in Europe and North America, and it was during this period that the basic turbine technology in use today was developed.

With the penetration of subsidized grid electricity into remote rural areas, there was a steady trend away from small hydro from the 1930s until the 1970s. Even rural areas without mains electricity generally found it cheaper and easier to install diesel generators than to bother with the complications of installing hydroelectric systems. In this context, it should be noted that, although turbine technology has not changed much, the control systems in use for SHP systems until recently were costly, tended to be troublesome to maintain, and often failed to keep the system functioning within the voltage and frequency limits normally expected for running modern electrical appliances.

The first oil crisis of 1973 was a major catalyst in prompting developed and developing countries alike to look to their indigenous energy resources for electricity generation. Both grid electricity and oil-based fuels became increasingly expensive in most countries up until the mid-1980s. Governments continued to encourage the development of small hydro resources, and there was a revival of the industry, with a few new turbine manufacturers appearing. However, the fall in conventional fuel prices in real terms through the 1980s and into the early 1990s, and the subsequent 'dash for gas', has tended to cause interest in developing hydro systems to slow down once more.

Liberalization of the electricity industries worldwide has tended to cause a greater development of small and embedded generating capacity by independent power producers (IPPs), but so far the inherent complexity of planning and developing small hydro systems has tended to discourage the large-scale take-up of SHP by IPPs.[10] An exception is in China, where it is expected that some 20,000 MW of SHP will be installed in the same period (1993–2009) needed to build the world's largest hydro plant, the 18,000 MW installation at Three Gorges. In other words, SHP capacity can be increased more rapidly than that of large hydro; furthermore, the lead time for SHP projects is months rather than decades, and they do not have a major environmental impact.

3.2.3 Present state-of-the-art

Hydro power has been technically feasible for decades, and given a favourable site it can be economically attractive, sometimes even offering the least-cost method of generating electricity. Least-cost hydro is generally high-head hydro, since the higher the head, the less the flow of water required for a given power level, and so the smaller – and hence less costly – equipment that is needed. Therefore, in mountainous regions even quite small streams, if used at high heads, can yield significant power levels at attractively low costs. Norway, for example, produces some of the cheapest electricity in Europe from its numerous high-head hydro installations. However, high-head sites tend to be in areas of low population density where the local demand for electricity is relatively small, and long transmission distances to the main centres of population can often nullify the low-cost advantages at the hydro plant busbar. Easily engineered high-head sites are also relatively rare, with most of the best ones in Europe already developed.

Low-head hydro sites (see Figure 3.2) are of course statistically much more common, and they also tend to be found in or near concentrations of population, where there is a demand for electricity. Unfortunately the economics of low-head sites tend to be less attractive. In fact, most low-head sites are at best marginally economically attractive

Figure 3.2 Small hydro power: 500 kW Chinese-manufactured Francis turbine installed in Cuba.

compared with conventional fossil fuel power generation (if no allowance is made for the external 'added costs' of using fossil fuels), and for this reason many such sites exist but have yet to be exploited.

Paradoxically, under modern conventions for financial and economic appraisal, a new hydro installation tends to appear to produce rather expensive electricity, since the high upfront capital costs are usually written off only over 10 or 15 years – although such systems commonly last without major replacement costs for 50 years or more. In contrast, an older hydro site, where the capital investment has been completely written off, is usually extremely competitive economically as the only costs relate to the low O&M costs (many SHP systems these days are operated under automatic control without human supervision).

We therefore have situations such as that in the UK, where new hydro projects find it difficult to attract government funding because their unit energy costs are too high (typically in the order of €0.06–0.08/kWh), when calculated using the prevailing discount and amortized over only 15–20 years. Old hydro plants in the same country produce electricity at unit costs in the region of €0.01–0.02. In short, the institutional and financial framework in most countries does not favour the take-up of new hydro plants[4] even if old ones seem economically attractive.

Figure 3.3 Benson Weir on the Thames: typical of many low-head hydro opportunities in Europe.

3.2.4 Industry and employment

European manufacturers dominate the world market for SHP equipment. It is worth noting that, of 150 small turbine contracts (greater than 1 MW) awarded in 1991–92, throughout the world, only 21 were assigned to non-European firms.[3] Total employment from this sector in Europe has been estimated at 10,000 workplaces,[10] and turnover at present is of the order of €400 million per year.[11]

3.2.5 Distribution of small hydro power within Europe

Table 3.3 illustrates the status of small hydro in the 15 countries of the European Union.

Table 3.3 Developed small hydro in EU countries[12]

Country	No. of plants	Capacity MW	Planned
Austria	1700	620	100
Belgium	38	90	0
Denmark	N/A	10	0
Finland	204	365	n/a
France	1350	1600	7
Germany	6000	1300	100
Greece	10	42	88
Ireland	34	13	5
Italy	1510	2000	236
Luxembourg	13	22	0
Netherlands	14	37	15
Portugal	74	230	n/a
Spain	982	1414	70
Sweden	600	250	40
UK	110	95	4
TOTAL	12,621	8074	665

3.3 GOALS FOR RESEARCH, DEVELOPMENT AND DEMONSTRATION

3.3.1 Long-term goal (by 2005–2010)

SHP currently accounts for over 2% of all electricity in the EU, i.e. approximately 30 TWh/yr.[13] An extra 4000 MW or so of new SHP capacity could be realized by 2010 given a more favourable regulatory environment. The undeveloped hydro potential worldwide is very much larger, and obviously represents a huge export market for European industry, if Europe continues to maintain a lead in this field (see Figure 3.4).

3.3.2 Medium-term goal (by 2000–2005)

The medium-term goal is for 1000–2000 MW of new SHP capacity installed in the EU by 2005 (with a similar or greater capacity of exported European equipment).

3.3.3 Short-term goals (immediate)

There is some scope for co-ordination of efforts to develop more cost-effective low and ultra-low head SHP systems, but the main, general needs are considered to be:[4]

• establishment of definitive data on existing installations and further potential as an aid to setting realistic targets for future developments

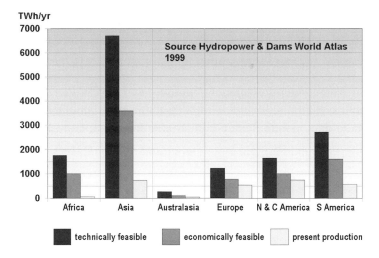

Figure 3.4 Hydro potential compared with present production for different regions.

- definitive evaluation of the real environmental impact of SHP compared with fossil fuel and nuclear power plants
- development of standard specification for design and installation 'packages', for export to developing countries
- harmonization of equipment standards at the European and at the international level.

3.3.4 R&D roadmap

The main area where the promotion of SHP take-up could be improved is non-technical – namely the institutional and economic framework for SHP projects. Gradually obtaining a more favourable infrastructure for the technology could be achieved by finding methods of finance and arrangements for electricity procurement that are more compatible with SHP. Further possible approaches include streamlining and simplifying the permission process for water abstraction from European rivers, and improving procedures for the sale of electricity to the grids.

The main technical thrust for SHP development is to improve the cost-effectiveness of the technology for use on the more common low-head sites. High-head hydro technology is generally fairly well developed, although the latest micro hydro technology, using electronic control systems, remains rare in developing countries, where it could find its largest markets. Only by improving cost-effectiveness can sites previously considered on the margin of being economic become attractive enough for widespread exploitation.

There are four main technical areas where research can assist the future dissemination of SHP technology, not only in Europe but worldwide, as indicated in the headings of the four sections that follow.[14]

3.3.5 Data collection on resource and on existing installations

Development of standardized methodologies for resource assessment and for evaluation of installations will act as an aid to setting more precise and realistic targets for future development.

3.3.6 Technical RD&D

To these ends, a number of approaches have been proposed in recent years and are beginning to be followed up to good effect.

3.3.6.1 General improvements

- **Standardization.** Overheads can be cut by reducing the need for specially designed, site-specific systems, although the potential for improvements in this area is relatively small, as modern computer-based design and NC manufacturing can allow rapid modification to cater for site variations.
- **Refurbishment of old sites.** This is one of the most promising and cost-effective ways to increase the European hydro generating capacity, as many thousands of old sites developed in the early part of the century have been abandoned (particularly in Eastern Europe) and many can be readily refurbished with modern equipment at marginal cost. For example, there are more than 3000 obsolete or abandoned hydro plants just in the eastern part of Germany (the former GDR).[15]

3.3.6.2 Improved turbines

- **Use of variable-speed turbines at low heads.** Recent developments in power electronics allow the possibility of turbines that do not need to run at synchronous speed, so that the turbine can be constantly regulated to run at an optimum speed. This permits much simpler (and less costly) turbines (such as fixed pitch propeller turbines) to be used without serious loss of efficiency, compared with the conventional single or double regulated Kaplan turbines normally used. Savings are in the order of 10%.
- **Use of submersible turbo-generators and siphon turbines.** These use technology similar to submersible electric pumps, have low maintenance requirements and can eliminate the need for a power house, or any direct supervision by an operator, thereby avoiding significant cost.
- **Use of new materials.** Plastics, new anti-corrosion materials, bearings and seals all offer possibilities for more cost-effective turbines, penstock pipes, etc.
- **Computer optimization of small systems.** This permits more accurate sizing, with the aim of optimizing the financial return from a site.
- **Innovative turbines.** Various novel types of turbine have been developed recently, and work has been done on using mass-produced – and hence inexpensive – pumps as turbines.

- **Head enhancement.** Surplus flow of water can be used to create an extra suction at the draft tube on ultra-low-head hydro plants in order to artificially raise the effective head – this can lead to reduced efficiency but increased cost-effectiveness when sufficient surplus flow is available.
- **Water wheels and small turbines.** There is possible value in redeveloping water wheels for small-scale applications on ultra-low-head sites and small turbines (<10 kW) for use on small streams.

3.3.6.3 *Improved electrical and control systems*

- **Use of induction generators.** These are less costly than conventional alternators, although their use involves some complications. They are in fact becoming the norm for modern grid-connected systems.
- **Use of electronic control and telemetry.** This permits unattended operation yet allows remote monitoring of the system to detect any incipient faults e.g. potential bearing failures, vibration, cavitation, etc.
- **Compact multi-pole generators.** There is the possibility of developing compact multi-pole generators capable of generating at low speeds to avoid the need for costly speed-increasers.

3.3.6.4 *Improved civil works*

- **Use of existing flood-control civil works.** With low-head systems in particular, the civil works often represent the greatest cost, but many weirs already exist for flood control purposes on most well-developed European rivers. The majority of flood control weirs do not have hydro generating equipment, yet a large proportion of these could have such systems added. Here the cost is reduced to just the marginal cost of fitting a hydroelectric system to an existing weir. It has been estimated that there is in the order of 3 GW of untapped but potentially economic SHP capacity of this kind on existing weirs in the EU.[16]
- **Inflatable weirs.** These are an increasingly popular technique for raising and stabilizing the head on sites with otherwise very low heads, which can of course increase the power and energy capture pro rata with the percentage increase in head.
- **Avoiding the use of cofferdams during installation.** Various system concepts and installation techniques can avoid or minimize the use of cofferdams, which are generally a particularly costly element of a hydro system installation.

3.3.6.5 *Improvements in ancillary equipment*

- **Simplification and improvement of trashracks.** Trashracks and trash removal can in many cases represent a major cost element in an SHP plant; various innovations, such as submerged and/or self-cleaning trashracks or self-flushing intakes, are being developed to mitigate this problem area.

- **Improved techniques to avoid interference or damage to fish.** This is an important issue in many rivers; one of the most common objections to new hydro systems is that they may harm or interfere with fish. Innovative forms of screening and fish ladder promise more cost-effective solutions to this difficulty.
- **Aeration.** Siphon turbines may be used to improve water quality by artificial aeration of water passing through the system.

3.3.7 Development of standard specifications for design and installation packages and harmonization of engineering standards

One of the reasons for the high costs of present generation SHP systems is that each one tends to be a unique design exercise, even if reasonably standard components are used. There is much scope for developing standard installation packages to suit reasonably common siting situations, so as to reduce the design overheads per site. There is also a need to improve the harmonization of SHP specifications, as there are at present numerous variations with which the technology may be presented or promoted. Furthermore, there is a major difficulty for potential users in evaluating the results they might expect when they seek to assess different options.

In the short term, one of the most promising areas for an expansion of hydro capacity in Europe involves both adding low-head hydro plants to existing weirs and flood control structures, and rehabilitating old and abandoned hydro plants. There is scope for facilitating this by developing standard installation designs to fit the most common types of weir and obsolete plants, so that the design inputs involved in specifying such equipment are simplified and made more economical.

3.3.8 Evaluation of the real environmental effects

Even though SHP systems are an environmentally clean method of power generation, there are still misapprehensions about them in many cases. False perceptions often arise from the adverse environmental effects widely publicized in connection with large hydro projects. But there are also a number of genuine negative environmental factors associated with SHP, including issues such as noise, aeration, fish damage, trash collection and disposal, residual river flow parallel to the hydro plant, and so forth, factors which can be made more or less severe depending on the installation design. It would be advantageous to explore these in detail, to produce definitive and broadly acceptable conclusions of use to assist both designers and planners in the future approval of SHP installations.

3.4 ROADS TO A STRONGER MARKET

Small hydro offers one of the most practical and immediately realizable routes to expanding the use of renewable energy resources in Europe, at the same time boosting European exports by strengthening the technically advanced European small hydro manufacturing industry. Yet it is handicapped more than most renewables by an unfavourable regulatory framework, and official support for the technology is generally limited. This is perhaps due partly to the fragmented nature of the industry and the involvement, mainly, of SME manufacturers with limited lobbying and self-promotional capacity, as compared with activists in other renewable energy fields. Hence the prerequisites for the future expansion of this industry are the need to raise official awareness of the benefits of small hydro, and the need to develop a more objective view of the true environmental impact of small hydro projects. This must be an essential component of any strategy to develop the use of hydro power in Europe.

Other factors in raising the public profile of small hydro relate to the development of effective and realistic standards for meeting requirements to minimize any environmental problems. These will need to avoid loading SHP developers with the unfair or onerous cost burdens that can result from unnecessarily stringent conditions. There is also a need for the tariff rates on offer to SHP owners and developers for the purchase of electricity to reflect the low environmental impact and high potential load factors from SHP, in contrast to those available to large hydro or conventional fossil fuel generation.

Hence the initial strategic requirement is for small hydro to feature more prominently in national and EC energy planning, with a view to its stronger encouragement. The regulatory and financial framework for SHP needs to be considered and improved and, where possible, harmonized across the EU.

Lastly, contrary to popular mythology, the fact that hydro represents a long-established and mature technology does not mean there is no further room for technical development. A resurgence of SHP development in Europe needs to be backed by technical improvements, designed to meet current and future requirements such as those suggested earlier. Therefore, SHP should feature as a valid area for R&D support in national and international programmes. Table 3.4 has been adapted from the

Table 3.4 Weighting scale for role of participants in development of small hydro in Europe

Key participants	Research	Development	Demonstration	Dissemination	Market penetration
Research centres (public)	1	1	2	1	–
Industrial/professional associations	–	–	–	3	3
Financiers	–	1	1/2	–	3
Utilities	2	2	3	3	3
Local authorities	–	–	2	2	2
Regional authorities	1	1	3	3	2
National authorities	3	3	2	2	2
EU institutions	1	1	3	3	–
Consumer associations	–	–	1/2	2	3
Civil society (NGOs, environmentalists, etc.)	–	–	1	3	2
Media	1	1	1	3	2
Manufacturers	2	2	3	3	3
Consultants	1	2	2/3	1	2
Installers/developers	–	–	–	2/3	3
Training/education centres	–	–	1/2	2/3	–

(1 = low, 2 = medium, 3 = high) (adapted from ref. 4)

recommendations of the expert group at the Madrid Conference for 'An Action Plan for Renewable Energy Sources in Europe',[4] and indicates the possible weighting of the roles of key participants in furthering the successful development of SHP in Europe. It shows that while there is a continuing need for R&D, the main thrust needs to be in the areas of demonstration, dissemination and commercial development to ensure market penetration.

REFERENCES

1   Energy Information Association, US Department of Energy.
2   *Layman's Guidebook on How to Develop a Small Hydro Site.* European Small Hydro Association and Commission of the European Communities, DG XVII, Brussels, 1994.
3   T T P Tung and K J Bennett, 'Small scale hydro activities of the IEA Hydropower Programme', *Hydropower and Dams*, September 1995.
4   *An Action Plan for Renewable Energy Sources in Europe.* Working Group Report on Small Hydro, Madrid Conference, 16–18 March 1994.
5   According to studies by the Spanish Institute for Diversification and Conservation of Energy (IDAE) and quoted in G A Babalis, 'The status of small hydropower in Europe', *Hydropower and Dams*, September 1994.
6   D L P Strange, Draft small hydro chapter, World Energy Conference Committee on Renewable Energy Resources, Opportunities and Constraints, 1990–2020. World Energy Conference, August 1992.
7   'Renewable energy sources: small-hydro sector', *Proceedings of Conference on Improving Market Penetration for New Energy Technologies: Prospects for Pre-Competitive Support*, EC DGXVII, October 1996.
8   *New and Renewable Energy: Prospects for the UK; Supporting Analysis.* ETSU R-122 for the DTI, Section on Hydro Power, April 1999.
9   *The European Renewable Energy Study*, ALTENER Programme, CEC DGXVII, 1994.
10 E A Maurer, 'New markets for small scale hydro', *CADDET Renewable Energy Newsletter*, ETSU, Harwell, UK, 1999.
11 G A Babalis, 'The status of small hydropower in Europe', *Hydropower and Dams*, September 1994.
12 World Atlas & Industry Guide – *International Journal of Hydropower and Dams*, 1998.
13 K V Rao and E M Gosschalk, 'The financial case for hydropower', *International Water Power and Dam Construction*, October 1995.
14 *Energy in Europe: Annual Energy Review, Special Issue 1993.* Commission of the European Communities, DG for Energy (DG XVII), Brussels, 1993; quoted in reference 4 above.
15 IT-Power, Stroomlijn & Karlsruhe University, *Technical and Resource Assessment of Low Head Hydropower in Europe*, Joule Project JOU2-CT93-0415, 1995.
16 Based on references 4 and 9 above, but with additions from various sources including personal communication from two hydro turbine manufacturers (1995).

4. ENERGY EFFICIENT SOLAR BUILDINGS

C F Reinhart (Fraunhofer Institute for Solar Energy Systems)

4.1 INTRODUCTION

Buildings fulfil multiple purposes. They provide shelter, and try to create adequate working and living conditions for their inhabitants. Apart from these functional aspects, buildings also serve as a means of cultural identification and social representation. To satisfy all these diverse expectations, financial, material and energy resources are required in their construction and maintenance.

Despite the diversity of individual lifestyles in the EU member states, about a third or more of the total final energy use in these countries is accounted for by heating, cooling and lighting buildings. In 1996, the associated costs in the domestic and tertiary sector corresponded to roughly 4% of the GDP of the EU (the number has been estimated from the energy demand, weighted with the European mean costs for oil, gas, coal and electricity[1]). Figure 4.1 shows that the annual energy-related carbon emissions per capita vary from 0.5 tonnes CO_2 for Portugal to 3.8 tonnes CO_2 in Luxembourg. The

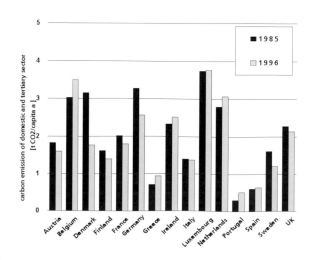

Figure 4.1 Energy-related carbon emission in the domestic and tertiary sector in the EU member states (ESAP sa 1998). The tertiary sector comprises all non-industrial, commercial activities.

levels have been falling since 1985 in many highly industrialized countries like Denmark and Germany, while countries with expanding industrial facilities like Greece and Portugal exhibited rising carbon emissions, due mainly to the penetration of air conditioning equipment.[2] Traditionally, a country's consumption figures have been determined by living standards, economic growth rates and energy prices, but recent developments in several EU member states show that emission levels can be decoupled from economic output and, to a certain degree, from climatic boundary conditions. Therefore, the development in Portugal and Greece does not imply that a minimum carbon emission level is principally necessary for economic activity. The future carbon emissions of all EU countries could fall well below the lowest levels found in Figure 4.1 if more sustainable building practices were fostered. In this context, it is important to realize that the contribution of passive solar energy to the total energy demand for space heating (solar gains, internal gains and active heating) is already substantial in buildings across Europe. Table 4.1 shows the individual level for several EU member states.[3]

Table 4.1 Solar fraction is defined as the contribution of passive solar gains used for heating to the total space heating demand in the heating season

Country	Solar fraction
Norway	10%
Germany	13%
Belgium	12%
Finland	15%
Greece	12%
UK	15%
Ireland	12%

Apart from that, increasing active solar contributions arise from the exponential development of the European collector market (Figures 4.2, 4.3).

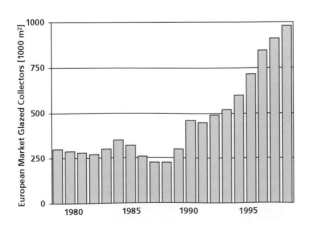

Figure 4.2 Glazed collector market in Europe 1980–1999.[4]

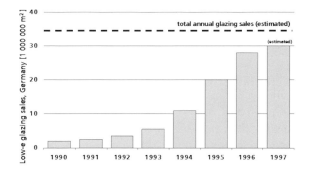

Figure 4.3 Development of low-e glazing sales in Germany. As of 1996, sales seem to have realised their maximum potential. Low-e windows now hold a market share in Germany of about 90%. (© Interpane)

To reduce the high carbon emissions in the building sector, a new class of buildings has been realized in the past few decades. These **solar buildings** or **lean buildings** are in harmony with their given climatic boundary conditions, and exploit naturally available energy sources in order to provide increased visual and thermal comfort for their inhabitants, while reducing the energy demand (examples can be found in reference 5). Solar buildings rely on both traditional, energy-conscious architectural practices and very recent developments that supply architects and building engineers with an ever-increasing catalogue of components and concepts. These include advanced windows and better insulation materials, solar cells and collectors, heat pumps, daylighting, integrated lighting and shading control systems, free nocturnal air cooling, preheaters (such as air heat recovery and air-to-ground heat exchangers for preheating in winter and cooling in summer) and phase-change materials, to name but a few examples. Some concepts work because of their simplicity, others because of their technical finesse. New, powerful simulation methods allow architects and engineers to model the interplay between these components in the design phase of a building (Figure 4.4). It is worthwhile to mention that, in the age of powerful, quasi climate-independent

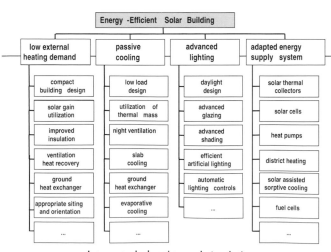

Figure 4.4 The architect and the HVAC engineers have a catalogue of components and concepts at their disposal to ensure that the energy concept of a building forms an integral part of the whole building design.

HVAC (heating, ventilation and air-conditioning) systems and falling energy prices, the construction of a lean building marks a conscious decision to find individual solutions, where conventional practices lack a sustainable dimension.

To realize the ambitious goal of constructing a lean building under economical boundary conditions, the design process requires more thorough planning than for a conventional building. Ideally, the building owner at first formulates a catalogue of requirements and issues for the future building. Certain weights should be assigned to the different items in the catalogue to reflect personal preferences, the available economical resources, the anticipated working conditions for the end users, and the sustainability of the resulting building. The composition of the design team should reflect these preferences so that an **integrated design process** can be initiated, in which the interrelations between the building and the HVAC system design can be addressed and possibly exploited (Figure 4.4). For instance, the extra costs created in the planning phase can be counterbalanced by reduced initial investments or lower future operation costs: i.e. costs are shifted from the investment into the planning phase. In the event that the financial situation of the building owner does not allow the implementation of various energy-saving components at the time of construction, flexibility for future improvements can be realized at little extra cost. Small changes in the layout of the electrical system and pipes in a new building can facilitate the later installation of photovoltaics, connection to a solar district heating system or the installation of a heat pump.

Innovative solar buildings follow a holistic approach, incorporating energy efficiency measures and an advanced control of the incoming solar gains, as well as solar assisted energy supply systems.[6] Alternative measures are assigned different priorities according to their potential for avoiding or reducing energy consumption and costs. Figure 4.5 suggests that while the overall design of a building should certainly not be cost-oriented alone, energy saving measures should follow a balanced **least cost planning** (LCP) approach to reach a desired total energy demand. Basic energy efficiency measures such as low-energy electrical appliances and improved envelope insulation tend to have the quickest monetary return, followed by efficient windows and an increased use of daylight. Solar assisted energy supply systems – like collectors and photovoltaics, for thermal and electrical energy respectively – are of increasing importance, as their

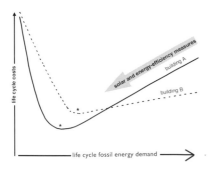

Figure 4.5 The energy concept of a lean building should follow a least cost planning (LCP) approach. The curves for two example buildings show that the total fossil energy demands and associated investment and running costs vary for individual buildings because of different user profiles, climatic boundary conditions and building types. Nevertheless, a balanced mixture of solar and energy saving measures reduces both costs and fossil energy demands, until a building-specific minimum is reached (*). The minimum corresponds to the energy concept of a lean, energy-efficient solar building.

Figure 4.6 Zero-energy house in Zandvoort, the Netherlands; integrated PV and solar thermal collectors, together with a high level of insulation, give this project a special appearance. The objective of the project is to demonstrate the feasibility of a self-supporting house with high living standards. (© R Schropp)

prices keep falling and their contribution to the total energy demand of the building rises. The obvious limitation of these energy systems is that their output is available only temporarily, so that they require either some means of storage, or a cogeneration power system. Example systems, which are elaborated upon below, include small heat pumps, biomass and hot water tanks. Fuel cells will also be available in the near future, probably at sufficiently affordable prices. If available, the local power grid is a low-cost storage system for electricity generated by photovoltaics. In this way, new solar buildings or districts can become net producers of electricity, if properly integrated into the demand-side management of the grid.

The energy demand of an individual building depends on its microscopic and macroscopic climatic boundary conditions. The former are extremely site-dependent, and include shading due to surrounding objects, and climatic peculiarities such as nearby natural waters. Macroscopic properties describe the regional climate including solar radiation data, outdoor temperature and precipitation. On-site energy autarky – although possible (Figure 4.6) – is usually neither sustainable nor economically viable as it involves disproportionately large energy supply and storage systems and high investment costs. As a solar building is a green product, total life-cycle energy balances, i.e. the energy necessary to construct, run and recycle a building, should also be taken into account.

Buildings require thermal and electrical energy. The former is used primarily for space heating and domestic hot water supply, while the latter is used for lighting, ventilation, cooling and user-specific electrical appliances. In order to quantify the overall energy sustainability of a building, the two energy types need to be compared with their corresponding primary energy conversion factors: i.e. the energy that needs to be generated in a power plant or provided through a gas pipeline to meet the on-site demand. As the generation of electrical energy in a power plant is accompanied by substantial conversion losses, costs as well as primary energy content of electrical power are higher than those of thermal energy. Table 4.2 reveals the high environmental impact associated with the generation of electrical energy and emphasizes the significance of measures that save electrical energy.

Table 4.2 Primary energy factors and CO_2 emissions according to TEMIS, 1999.[7] A conversion factor specifies how many kWh of primary energy are needed to provide 1 kWh of final energy in a building. The conversion numbers for electrical power are based on the German energy mix

	$MWh_{primary}$ per MWh_{final}	kg CO_2 per MWh_{final}
Natural gas, oil	1.1	210–290
Electrical power	2.9	640

As the degree of energy self-sufficiency of buildings rises, an important issue is how the remaining demand should be met in the future. Once a building has reached a certain, low energy standard, the options are either to employ further energy saving measures on-site, or to build solar networks to meet the remaining energy demand for a group of buildings. Solar district heating may increase in importance due to the large-scale potential for financial savings. In combination with a gas or biomass boiler, these systems can be financially competitive even today.[8] As for the single housing unit, the energy supply concept for a particular district is always a compromise, and a mixed cost and sustainability analysis is the best way of developing an appropriate solution.

4.2 PRESENT SITUATION

This section sketches the present state of the European building stock, describes recent developments in construction practices, and presents new building components. To gain more insight into the dynamics within the building stock, three building categories that require different energy saving measures and solar technologies are discussed separately, namely:

- new residential buildings
- new non-residential buildings
- the existing building stock.

The different components and concepts described in this section are ordered by relevance to the building category. This does not imply that they are exclusively used in that specific building type.

4.2.1 Residential buildings

The energy demand of residential buildings in northern and central European climates is dominated by the heating demand, which can be reduced dramatically by better insulation techniques, improved windows and optimized heating systems. In southern Europe, the cooling demand can be reduced in a similar way. A compact design reduces energy exchanges over the surfaces, but also reduces the availability of daylight and solar gains in the building's interior. Accordingly, the size and position of windows should take into account the orientation of their corresponding facade: reduced window apertures to the north and large apertures to the south have a positive influence on the energy balance in the heating season. Apart from the type and quantity of the insulation materials, high construction quality is essential to avoid heat bridges and create a thermal envelope with a low overall heat transfer coefficient. This construction quality will only become a standard if the training of construction site foremen adapts to the changing requirements.

In Austria, Germany, Sweden and Switzerland particularly, a new generation of low energy housing is starting to set a certain standard among new residential buildings (Figure 4.7). The heating demand of these buildings lies as low as 15–30 kWh/m^2 per year and examples have been realized at approximately 10–20% above initial construction costs. (The mean heating demand in the German building stock lies at around 220 kWh/m^2 per year.) Prominent construction features are improved insulation of the building's envelope, combined with advanced windows and controlled ventilation with air heat recovery. Solar domestic hot water is a standard in these buildings since, as the energy use for space heating falls, the energy demand for hot water is gaining importance. The trigger for positive development in these countries has been stricter building codes and a growing environmental awareness, which has in turn created rising demand and falls in price for the energy-saving components utilized.

Figure 4.7 Low energy buildings in Salzburg, Austria. 60 apartment units with solar assisted heating and domestic hot water (solar fraction 35%, total collector area 400 m^2, hot water storage tank 100 m^3); the remaining heating demand corresponds to 50 kWh/m^2 per year (architect: W Reinberg). (© AEE-Gleisdorf)

Following the reductions in energy use in new single-family dwellings, the focus of research is shifting towards low energy apartment buildings. Conceptually, apartment buildings are better suited for energy saving and solar concepts than single family houses: they tend to be more compact than single family houses and command reduced prices due to the economy of scale. Furthermore, they accommodate the majority of the population in many western societies. A barrier for a wider market penetration of solar apartment buildings is the landlord–tenant problem – while the costs of energy-saving measures are paid for by the landlord, the tenant benefits from reduced operation costs. The issue might be resolved in the future when high thermal comfort conditions and low operation costs are appreciated and indeed expected by a growing number of people.

4.2.1.1 Windows – solar heating and daylight

Windows are multi-purpose architectural elements that provide visual contact with the outside, and serve as sound reduction, security and weather protection devices. As windows are the components with the highest energy transfer in the building envelope, it is necessary to reduce unwanted heat transfer through them in both directions, and to manage the incoming solar gains according to the building's needs at a given time. The amount of thermal losses is characterized by the U-value (heat loss coefficient) while the admittance of solar radiation is described by the total solar energy transmittance, or g-value. The g-value is the sum of the solar transmittance and a variable for heat gains resulting from sunlight absorbed in the glazing unit. The crucial parameter for daylighting is the visual transmittance, the transmittance of the visible part of the solar spectrum.

Independent of the building site, a low overall U-value for the window unit is desirable, as it reduces energy flows between the continuously changing ambient temperature and the indoor temperature, which exhibits lower daily and seasonal variations. Various design measures have been developed to reduce these heat losses:

- **Low-e coatings.** A crucial technical innovation to limit the thermal radiation exchange between two glazing units is to deposit a selective coating on a surface bordering the gas-filled gap between two glass panes, which acts as a mirror to the infrared radiation: i.e. visible light is still admitted, while infrared radiation exchange between the glass panes is impeded. Different coatings allow for an individual setting of the ratio of solar-to-visible transmittance. The thin coatings usually consist of multiple layers including a noble metal or doped semiconductor oxides. Secondary solar gains from the absorption of solar radiation in the glazing can be admitted into the room, or excluded if the building is situated in a cooling dominated climate, through the choice of which glazed surface is coated. In this way, a permanent solar gain control is realized.
- **Triple glazings.** Another way to reduce heat losses is to integrate a third glazing. Combined with a low-e coating, this can lead to centre of glazing U-values as low

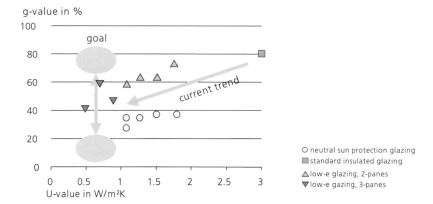

g-value in %

O neutral sun protection glazing
■ standard insulated glazing
△ low-e glazing, 2-panes
▽ low-e gazing, 3-panes

U-value in W/m²K

Figure 4.8 Development of the total solar transmittance (g-value) with respect to the heat transfer coefficient (U-value) in modern glazing units; all glazing units shown are commercially available and the U-values refer to the centre of glazing. Innovative technologies have led to reduced heat losses (falling U-values) through windows. This positive development has been accompanied by a slight fall in solar transmittances as additional glazings or coatings necessarily reduce the overall transmittance of the window unit. The long-term goal is flexible control of the solar gains via switchable g-values.

as 0.5 W/m²K (Figure 4.8). The drawbacks of triple glazing are larger window frames, needed to support the additional weight, and a reduced g-value of the glazing unit. A recent approach, already commercially available, is the use of glazing with reduced iron oxide content. In this way the visual transmittance of the window rises from 60% to 75%, which corresponds to the visual transmittance of double glazing.

- **Noble gas fillings or evacuation.** To limit convection between two glazings, it is possible either to use a noble gas in the enclosed space or to evacuate it. In both cases, the long-term stability of the sealing edge is crucial. These window units exhibit overall U-values below 2 W/m²K. Both technologies are already employed in various commercially available products.

- **Transparent insulation materials.** Another means of reducing convection is so-called transparent insulating materials. By introducing transparent open-celled structures made out of plastic or glass, the interior space between two glazings is divided into small cavities. As direct view through the resulting window unit is not possible, the application of this material is confined to non-view areas such as skylights. Transparent insulation materials are also used in combination with walls. In this latter application, the transparent insulation functions as a radiation trap for the incident solar radiation. In the winter, this leads to heated wall surfaces, which provide high thermal comfort; in the summer, an additional shading device is usually necessary to avoid overheating.

- **Frame and edge seal technology.** As the thermal performance of the glazing unit improves, the surrounding construction must become compatible, to ensure that the entire window unit functions in all respects. The U-values of the spacer and the geometry and material of the frame are decisive factors. Following the development of the glazing unit as outlined above, innovative spacers should exhibit thermal improvements over metal spacers, be gas-tight with long-term integrity, and incorporate drying agents. Several alternative spacers are being examined as a replacement for conventional spacer materials such as aluminium and stainless steel. The frame is usually made out of wood, PVC or aluminium, and combinations with fibreglass and plastic foam thermal bridges are being tested to reach window units with an overall U-value of 0.7–0.8 W/m²K.

Figure 4.8 sketches the effect of different measures on the U-value and g-value of various glazing units. All examples shown are commercially available. The associated costs and the availability of these innovative window technologies vary markedly across different EU member states. While the current trend is oriented mainly towards lower heat loss coefficients, the figure suggests that the long-term goal is flexible control of the incoming solar gains (see Sections 4.2.2.3 and 4.3.1).

4.2.1.2 Collector and storage systems – solar domestic hot water and solar assisted space heating

As explained above, residential low energy buildings usually feature solar collectors either to preheat domestic hot water and/or to cover a fraction of the space heating demand. As for the whole building, an integral approach should be used to identify the most appropriate system layout. A collector yields yearly solar gains of 300–550 kWh/m², depending on the local climate and the average temperature in the collector circuit. The average system dimensions for a single-family housing unit differ for the Mediterranean and Northern Europe. A typical solar heating system found in the South has a collector area of 2–4 m² with a 100–200 litre storage system. The back-up heating is usually provided electrically. Such a system covers roughly 90% of the annual energy demand for domestic hot water with solar energy, and costs €200–300 per m² collector area.[9] A typical Northern solar system for domestic hot water has a collector area of 4–6 m², a gas boiler and a 200–300 litre hot water tank. The solar contribution to the heating of domestic hot water demand lies at about 60%. The system costs range between €600 and €1000 per m². They tend to be higher than in the South because the collectors usually feature a selective absorber and a transparent cover. The market penetration of solar heating varies considerably across the EU. Austria and Greece exhibit the highest installed collector areas per capita, 0.028 and 0.012 m² respectively. Reasons for purchasing collector systems tend to be cost-oriented in Mediterranean countries and environmentally motivated in the North. Figure 4.9 shows the price development per m² of collector area with respect to the cumulative collector area in Germany.

To further reduce system costs, large-scale solar applications (>100 m²) have been investigated in projects in Austria, Denmark, Germany, Greece, Italy, Sweden and the Netherlands.[8] In a **solar district heating system**, a large collector area is connected to

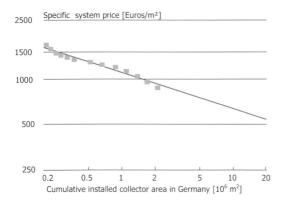

Figure 4.9 Development of average specific system prices for solar heating in Germany. The data are based on cost developments between 1985 and 1997.[10] Growing production numbers and intensive material research seem to have led to continuously lower prices for collectors.

a district heating network, which distributes hot water among the participating households. Such projects profit from reduced installation and material costs, as prefabricated roof-module collectors (>20 m²) can be employed in replacing the regular roofing material. Solar district heating systems with a diurnal storage tank have been realized at total system costs as low as €200–600 per m² of collector area. Solar district heating systems with a seasonal storage facility (volume 1500 m³) can cover more than 50% of the overall heating demand, i.e. space heating *and* domestic hot water. Total system costs are about twice as high as for a diurnal storage system. The development of cheap, large storage systems is an area of intensive research, as the seasonal storage of solar energy greatly reduces annual carbon emissions, especially if the system is complemented by a gas or biomass cogeneration system. In the long term, 80% of thermal energy used today could be saved or delivered by such systems. If a constant source of waste heat from a nearby industrial site is available, then district heating systems are a financially viable option.

The remaining barriers to a wider market penetration of solar district heating systems are of a non-technical nature. Municipalities have considerable interest in creating residential areas with a low environmental impact. The benefits range from environmental awareness to publicity and high-quality living conditions for the inhabitants. On the other hand, the financial resources of most municipalities are extremely limited, and 'solar districts' require higher initial investment costs than conventional systems. Even if these costs are amortized by reduced operation costs, the public financing system of various EU countries does not allow funds to be shifted from maintenance to investment budgets. One challenge for the future will be to restructure these financial systems, and to identify partners who can help municipalities realize large-scale solar district heating systems. Banks that raise money in tax-free green funds or third-party financing might be suitable sources of necessary funding. The optimum cost-effectiveness and widespread implementation of solar heating will depend strongly on national energy tax strategies, which define the boundary conditions for such financing schemes.

4.2.1.3 System integration for domestic hot water and heating

The earlier sections have shown that the trend in low energy housing is to provide a mixed supply system for domestic hot water and even space heating based on a collector, a thermal storage system and a cogeneration system, whether a gas, oil or biomass boiler or even a fuel cell. The interaction between the various components is complex, and system losses resulting from unsuitable installation or control strategies may seriously reduce the performance of the whole system. To address this problem, a number of integrated energy supply systems have been developed:

- **Compact domestic hot water systems.** While in the 1980s the installation of a collector system was usually carried out by a local provider with a high degree of manual work, the market – especially in Austria – is now becoming dominated by larger enterprises that increasingly offer integrated systems. The advantage of these innovative products is that the storage tank, the collector module, the cogeneration system and the control unit are provided by a single company. This bundling permits lower total system costs and a higher degree of prefabrication, which reduces on-site installation errors or incompatibilities of the various components, and also provides increased security for the client. In these system kits the cogeneration heating unit is often an integral part of the water storage tank, to reduce heat losses.
- **Solar combi-systems.** The demand for solar heating systems that cover not only domestic hot water preparation but also space heating is rising. In several countries, such as Austria, Denmark, Germany and Switzerland, such systems already hold a significant market-share (Figure 4.10). Such combined systems, also termed solar **combi-systems**, resemble domestic hot water systems as far as they also feature collectors and a storage device. The main difference is that solar combi-systems tend to exhibit larger collector areas and need to supply two different heat consumers

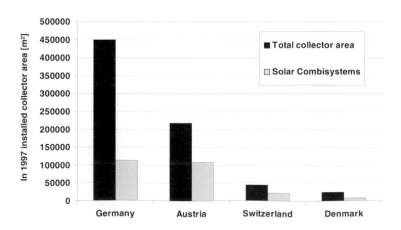

Figure 4.10 Installed collector areas and share of collectors for solar combi-systems in selected countries, 1997. (© AEE)

(a) *(b)*

Figure 4.11 Solar combi-systems. (a) Low energy building complex with six terraced houses and an office building in Gleisdorf, Austria, with an annual space heating demand of 20 kWh/m² per year. The total collector area is 230 m² with a storage volume of 1 m³ for domestic hot water and 14 m³ for space heating. The auxiliary heater is a Pellet Boiler (architect: W Reinberg). (b) Hotel at 2000 m altitude near Silvretta, Austria. The collector area is 60 m² with a storage volume of 1 m³ and 14 m³. The backup heating system runs on electricity (architects: Much-Untertrivaller & Hörburger). (© AEE)

with different demand profiles (Figure 4.11). This leads to a system with a storage tank, two supply sources – the solar collectors and the auxiliary energy source – and two heat consumers. The interaction between these five subsystems is complex and profoundly affects the overall energy performance of the whole system.

- **Advanced control algorithms.** Due to the emergence of integrated solar supply systems and powerful low-cost computing units, the control unit can be programmed with more advanced, so-called **match flow** algorithms as the interaction of the system components is better understood. The aim of matched flow techniques is to adapt the flow rate in the collector circuit according to user habits, the day of the week and seasonal changes. Innovative control units are fully integrated into the overall building technology, and rely on fuzzy logic and neural network algorithms to anticipate short-term ambient temperature and solar gain developments. This can reduce the operating time of the back-up system by about 10%.

  Recently developed operating networks enable remote control of building control systems, and the idea of outsourcing the maintenance of these systems is gaining momentum.

- **Waste air heat pumps.** The integration of a compression heat pump with a collector module, a water tank and a controlled air ventilation system is another way of covering the thermal energy demand of a building. In a low energy house with a heating demand below 15 kWh/m² per year, the waste air is almost sufficient as a heat reservoir for the heat pump to heat the incoming air. As electrical energy is necessary to run a heat pump, care has to be taken that the system's coefficient of performance is high. The performance depends mainly on the annual operation time of the cogeneration system, which is also usually driven electrically. The distinctiveness of this heating system is that it can be sustainable even though it is electrically driven.

4.2.2 Non-residential buildings

Non-residential buildings exhibit conspicuously different energy demands from those of residential buildings owing to different user expectations and occupation times. Figure 4.12 shows some recent trends in the European office-like building sector. Shown is the development since 1972 of the mean annual thermal and electrical energy demand per net floor area of new, non-residential buildings in the Swedish building sector.[11] Qualitatively, the data can be transferred to northern and central European building stock. The data reveal that Swedish office buildings have experienced a continuous decrease in heating demand following the development of stricter building codes. This increased efficiency in the thermal envelope has been accompanied by higher electrical energy demands. Two prominent examples for this development are the two monitored Swiss buildings[12] in the upper left corner of Figure 4.12, which combine uncommonly high electrical energy demands with an excellent thermal energy demand below 50 kWh/m² per year. Several circumstances support the present trend:

1 Most modern office buildings aim to create a time- and site-independent indoor climate that approaches a narrow, *trans*-global range for what are considered to be adequate working conditions.[13] As the tolerance range of the users becomes smaller, the need for powerful climatization equipment rises, together with the energy demand. The effect is pronounced if many workplaces are grouped together in large offices.

2 Commercial buildings communicate the corporate identity of the building inhabitants to the outside world, and place the buildings in the context of their surroundings. This important function of buildings has led to examples of buildings that ignore their climatic boundary conditions for the sake of a desired visual impression.

3 As modern office buildings exhibit rising internal load profiles due to the explosive increase in the use of electrically powered office equipment (such as PCs, printers

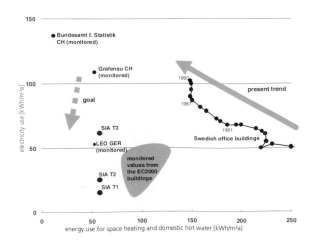

Figure 4.12 Trends in the European office-like building stock. The annual energy demand refers to the net heated floor area of the buildings. The data show monitored energy demands of single buildings or building groups except for SIA T1 T3, which are energy benchmarks specified by the Swiss Society of Engineers and Architects (SIA) for different commercial building categories. (© H. Hartwig, Technical University Munich, Germany)

and other technical equipment), the cooling loads start to dominate the overall energy use even in moderate climates. The peak of this development seems to have been reached as more energy-efficient appliances are finding their way into offices, e.g. LCD computer displays and fluorescent lamps. In Switzerland, the main motivation for purchasing these new appliances has been found often to be unrelated to their greater energy efficiency: lower energy demand tends merely to be a positive side effect. Electrically driven, cooling dominated HVAC systems and the growing use of artificial lighting increases further the overall CO_2 emissions caused by a building's energy use.

4   Furthermore, energy saving measures and building energy codes tend to concentrate on the thermal envelope, since the thermal energy demand dominated a building's total energy demand in the past. An exception to this is the Swiss SIA building code, which formulates standards for thermal energy, as well as electrical energy use for HVAC and lighting. The SIA is the Swiss Society of Engineers and Architects. Figure 4.12 shows the ambitious SIA energy benchmarks for office buildings of varying technical standards (T1 and T2) and with air conditioning facilities (T3).[14]

The diversity of the above issues highlights the complexity of the phenomena of rising electrical energy demands in commercial buildings. Point 1 is closely linked with the expectation and attitude users have towards their working environment, and how the climatization concept enables them to influence it. If the climatization system allows for only minimal seasonal and diurnal indoor temperature variations, the users become increasingly separated from their natural surroundings. This leads to possible health problems[15] and necessitates oversized technical equipment. Point 2 is directly related to the architect's understanding of the significance of energy within the overall building concept. Improvements could be realized by innovative educational colloquia. Points 3 and 4 address society and policymakers, and their awareness of electrical energy use.

Given that there is a consensus between a building owner and an architect that a commercial building should be both energy-conscious and have high comfort conditions, how far can energy demand be reduced in comparison with conventional buildings? Figure 4.12 shows some exemplary buildings like the Low Energy Office (LEO) in Cologne, Germany, constructed in 1995.[16] The office buildings from the Energy Comfort 2000 project (EC 2000) also point the way to reduced specific electrical energy use. EC 2000 was a demonstration project on energy-efficient non-domestic buildings funded by the European Commission.[17]

To realize such projects, intensive communication between the architect and the building engineers is necessary. Computer simulations are usually required, as commercial buildings tend to be larger than residential buildings and the interaction of the various energy flows is more complex. Their understanding becomes crucial if innovative daylight and climatization concepts are to be investigated. The energy flows in a commercial building are characterized by substantial internal loads and solar gains, as increasingly large glazings are a feature of modern architecture; consequently, solar gains in such buildings need to be controlled. Electrical energy savings through daylighting also have a high potential, since the periods of daylight availability and building occupation largely overlap; it is therefore possible to reduce artificial lighting loads. If internal loads and solar gains are sufficiently reduced, a **lean climatization concept** can be developed. The term **lean** signifies that the system is electrically efficient so that only a small amount of electrical power (for running fans and circulation pumps) is required to maintain comfortable indoor temperatures all year around. If a lean climatization concept alone is not capable of providing this, combination with a solar assisted sorptive cooling system is an option for maintaining a low electrical energy demand. It is important to mention that the success of an energy-conscious climatization concept requires the careful assessment of the microclimatic boundary conditions and working profiles of the future user. Significant modifications to either parameter, e.g. a changed shading situation or a higher user density, may seriously affect the functioning of such a concept, especially if free nocturnal ventilation is involved. Additional planning efforts can be balanced by reduced investment and running costs.

The construction costs of a lean commercial building are project specific. Nevertheless, Figure 4.13 shows that the investment costs of several exemplary large-scale lean commercial buildings (>1000 m^2) lie within the range of the average costs for office buildings of medium to high standard in Germany.[18] Care should be taken when transferring the figures given in Figure 4.13 to other countries, as the costs for energy-efficient measures are closely interwoven with regular construction costs. Apparently, the size and type of a building and the economic situation in the construction industry tend to have a greater influence on the final costs than peculiarities of the energy concept.

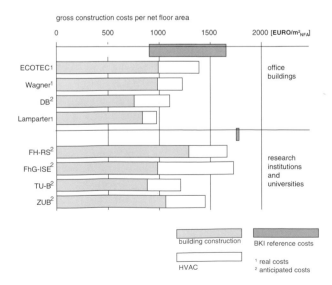

Figure 4.13 Investment costs, including tax, for the building construction and HVAC system for several exemplary buildings; planning and site costs are not included. The reference costs are published annually by the German 'centre for construction costs', and are based on mean German construction costs for a given building type.[18]

4.2.2.1 Visual and thermal comfort

The demand for high comfort conditions for the users of a building is omnipresent, even though there is comparatively little known on this issue. Thermal and visual comforts quantify a building user's experience of the lighting and temperature distribution in a workplace. Indoor environmental quality is a combination of acceptable indoor air quality together with satisfactory thermal, visual and acoustic comfort conditions. Personal comfort is difficult to measure, as it is highly subjective and depends on behavioural situations. Nevertheless, several thermal comfort equations have been formulated that yield the satisfaction of an ensemble of users, as a percentage, for a steady-state working situation. Recent research has included field studies to investigate adaptation to higher temperatures, and preliminary findings confirm that people are comfortable at much higher temperatures than expected if they can control their own lighting, heating and ventilation and have a clear view out of the window.[13] This implies that lean cooling concepts do not need to follow the static and narrow temperature levels for which conventional air conditioning systems strive.

Glare is more difficult to quantify or predict than thermal comfort. Several glare indices have been formulated, but the reliability of these for visual comfort predictions is still a subject of intensive research.

Inadequate building ventilation is the primary cause of poor indoor air quality, which may lead to illness, complaints and poorer performance at work. Ventilation problems have been exacerbated during recent years as the incoming flow of fresh air has been reduced in order to save energy. The elimination of the humidification or dehumidification systems and the early shutdown and late start-up of ventilation systems also contribute to poor indoor air quality. It is important that users' well-being remains the central performance indicator of a climatization system.

4.2.2.2 Computer simulations

Building simulation tools have been around for a long time. Some are used to describe physical effects in individual building components, while others aim to describe the building as a whole. Traditionally, separate simulation tools with different numerical algorithms have been necessary for thermal, lighting and ventilation, to model the energy flows due to these phenomena.

The simulation of thermal energy balances in a building has greatly advanced in recent years, and state-of-the-art programs like ESP-R[19], TRNSYS[20] and ENERGYPLUS (former BLAST and DOE2[21]) feature a continuously growing database for modelling standard and exotic system components, ranging from solar collectors to heat pumps.

Excellent daylight simulation tools like RADIANCE[22] and LIGHTSCAPE have existed for many years. Present research efforts aim to allow the simulation of innovative daylight elements, improve the available sky models, and evaluate the annual performance of a daylighting concept. A key problem here is to quantify the appearance of glare and visual comfort.

Programs to model the energy-related air flows in a building such as PHOENICS[23] are increasingly used by planning teams. The main difficulty for such programs is the high level of calculation power needed to simulate even straightforward room geometries and set boundary conditions such as ambient wind profiles. The problem is even more pronounced in an urban setting with complex aerodynamic conditions.

Recent developments aim to bundle different simulation tools into a single program in order to reduce time-consuming and error susceptible input, and facilitate maintenance of building models. Another advantage of these **integrated performance tools** is that they yield information for a variety of design aspects, ranging from indoor comfort to CO_2 levels and energy demands. These diverse criteria can be presented in a compact **integrated performance view**[19], as in Figure 4.14.

As the significance of computer tools in the design phase of a building increases, there is a growing need for internal consistency checks, and a thorough error analysis of the simulation results.

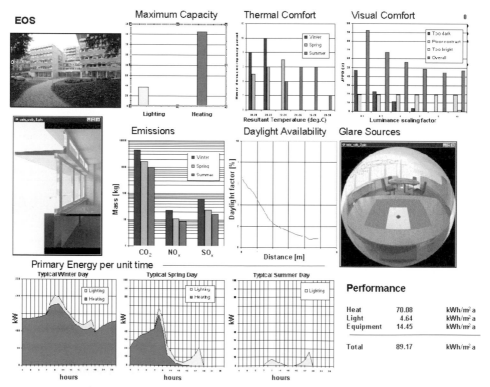

Figure 4.14 Integrated performance view (IPV) for the Headquarters of Énergie Ouest Suisse in Lausanne, Switzerland. The building features a central atrium and external light shelves. The IPV is a compact representational mode for various building parameters like maximum lighting and heating capacity, thermal and visual comfort, emission levels, daylight availability, glare sources and primary energy demand per unit time. The figure is meant only to provide the reader with a visual impression of an integrated performance view. No in-depth understanding of the single graphs is anticipated. (© S Citherlet, LESO, Switzerland)

4.2.2.3 Solar gain control

While reducing the U-value of a window unit is a common goal for architectural glazing in all climates, the requirements for the g-value and the visible transmittance are more complex. In cooling dominated climates, the purpose of transparent facade elements is to admit sufficient daylight and to allow visual contact with the outside. Solar gains need to be avoided most of the time. This is best achieved by glazing with a low g-value and a high visual transmittance. In temperate and northern climates, the situation becomes increasingly complex as solar gains are welcome in the winter and unwanted in the summer. This calls for versatile window technologies that try to address the seasonally varying demand. Traditionally, this task has been realized by a combination of a window unit with a shading and glare protection device. Outer lamellae and venetian blinds are widely used, are versatile, and exhibit high user acceptance. High-rises are exposed to high wind loads so that the shading systems need to be placed inside the window unit or the building. With this layout, there is a danger that the incident solar radiation is not fully reflected back out of the building. Exterior shading devices require regular cleaning and are susceptible to damage because of harsh weather conditions, which have a severe impact on the visual appearance of a facade.

Several innovative technologies aim to replace conventional shading devices. **Solar control** is the term that describes the deliberate, permanent or temporary reduction of the solar transmittance of a glazing to avoid glare and/or overheating. Multiple techniques are being investigated to limit and control the g-value of a window unit.[24,25] Some of these are permanent, and require combination with an additional shading device such as venetian blinds. More advanced, so-called **switchable** glazings feature a variable g-value so that a low U-value and flexible glare and solar gain protection can be provided by a single window unit. Electrochromic glazings, discussed below, are an example of switchable glazings that have already entered the market. Further technologies on the verge of entering the market – like photochromic, thermotropic, gasochromic and angular selective glazings – are described in Section 4.3.

- **Electrochromic glazing.** Electrochromic glazing is presently the only commercially available switchable glazing type.[26] The optically active layer of the glazing units is tungsten oxide and the switching signal is external. The switching is induced electrically, as the system functions like a large-area galvanic cell protected by two glazings (Figure 4.15(b)). The key components are two transparent conducting coatings through which an external voltage of about 3 V is temporarily induced across an ion storage layer, an electrolyte and the electrochromic layer. An external voltage forces Li^+ ions through the electrolyte into or out of the WO_3 layer, which turns blue or bleaches. In the absence of an external voltage, the electrochromic layer remains in either state due to the low electron transmissivity of the electrolyte (open circuit memory). The visual transmittance can be adjusted between 12% and 50%. For such glazing, switching time is a maximum of 12 minutes. Glazing sizes up to 2000 mm x 900 mm are available at system costs of €500–750 / m^2.

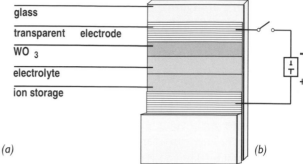

glass

transparent   electrode

WO $_3$

electrolyte

ion storage

(a)   (b)

Figure 4.15 (a) The headquarters of the Stadtsparkasse Dresden features a switchable facade with an electrochromic glazing (architects Bauer & Keller, Dresden, Germany); the left part of the glazing is in a coloured state (b) The physical principle of electrochromic glazing.

4.2.2.4 Daylighting and artificial lighting

Daylighting, the immediate exploitation of solar energy, forms an established part in the integral design process of a commercial building. It strives to optimize the availability of glare-free natural daylight to light the interior of a building. Natural daylight is extremely variable and cannot be stored, so the success of a daylighting concept for a specific site depends on its geographic position, the local climate, the shading situation due to surrounding objects and the overall building design. The benefits of a carefully planned daylighting concept range from enhanced visual comfort for the inhabitants to reduced use of artificial lighting. Two factors determine the success of a daylighting concept: the annual daylight availability in a given workplace environment, and how the incoming daylight is perceived by the users. The latter is highly dependent on the users' assigned tasks as well as their subjective and culturally motivated preferences.

The ultimately available quantity and quality of daylight in a building is decisively influenced at several design stages. The early design phase is crucial, as it is then that the distribution of the building masses on the site, the orientation of the building, the room depths and ceiling heights as well as the horizontal transparency of the building are defined. Later in the design process, the suitability of a facade for daylighting is determined by the position and size of apertures, the width of the window frames and the types of glazing utilized. Further important parameters are the photometrical properties of the ceiling and wall surfaces, the kinds of shading devices used, and finally the artificial lighting system. The latter should play the role of a back-up system for the available daylight.

Artificial lighting makes up about a third of the total electrical energy used in a building. An efficient lighting design can reduce this value by up to 70%. This can be achieved by introducing a control system that activates the lighting when needed, based on occupancy and indoor lighting control. Other savings can be realized by using lamps with higher efficiency lighting and advanced reflectors.

The use of daylight is closely linked with the solar gain control system. The goal is to admit sufficient daylight but avoid unwanted solar gains. This apparent contradiction can often be resolved, as it is usually only scattered, diffuse daylight that can be used for lighting. Direct sunlight is a source of heat and glare, and can only be used for lighting a workplace relying on computer use if it is redirected, e.g. via the ceiling. For this task, several daylight elements have been introduced. Their major advantage is that they offer the possibility of supplying such computer-reliant workplaces with glare-free daylight, while saving energy by reducing the operating time of the artificial lighting system. Figure 4.16 outlines the principle of a daylight element. As there is usually a surplus of daylight near the facade and too little daylight in the back of a room, a daylight element redistributes the incoming daylight. It is still unclear how successful daylight elements are at improving indoor comfort conditions or saving energy. As explained above, a high degree of uncertainty arises from the fact that the nature of visual comfort is not as yet sufficiently understood. The energy saving potential of a specific element depends on whether its visible transmittance under cloudy sky conditions (which are predominant in northern and central Europe) is comparable with a conventional glazing. Otherwise, benefits yielded on sunny days are counteracted on cloudy days. Examples of commercially available systems include two-component venetian blinds, shading devices that close from below, reflective structures, laser-cut panels and holograms. High system prices are still a barrier to wider market penetration.

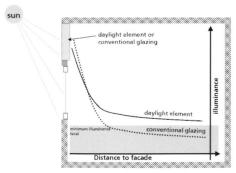

Figure 4.16 Principle of a daylight element: shown is the development of the indoor illuminance distribution with conventional glazing and a daylight element under a representative sky condition. In contrast to conventional glazing, a daylight element aims to 'even out' the indoor illuminance distribution so that the minimum illuminance level is maintained throughout the room without the use of artificial lighting.

4.2.2.5 Lean cooling concepts

A lean cooling concept lies at the heart of integrated design since it involves multiple design aspects. Microclimatic and user-specific boundary conditions must be carefully assessed. Internal and solar gains need to be controlled, and air flows within a building need to be predicted to avoid indoor temperatures above 26°C and maintain a high level of thermal comfort for the inhabitants. Internal loads can be reduced by purchasing energy-efficient devices and avoiding wasteful stand-by periods. As with the daylighting concept, suitable solar gain management is supported by a suitable facade design, facade orientation, window unit and shading device. The rising expectations for facades have led to the development of multi-purpose facade elements, which feature glare and heat protection elements as well as photovoltaics and automatically controlled openings for natural ventilation. Atria enhance the attractiveness of a building's interior and admit daylight deep into the building. However, they are also susceptible to excessive solar gains in the summer. The ventilation concept is designed around all of these influencing factors, its purpose being to provide fresh air at comfortable temperatures for the users. Heating and cooling can be realized with the air as carrier medium, via radiators, or by piping integrated into the building structure. Controlled ventilation is an efficient way to limit the air exchange to a suitable standard of hygiene (approximately 30 m³/h). A natural ventilation system is considerably cheaper than a mechanical ventilation system but requires the co-operation of the user. Two energy-conscious cooling concepts are presented below:

- **Free nocturnal ventilation.** A sunny summer day is characterized by high diurnal ambient temperature variations. In Freiburg, Germany, ambient temperatures can range from 14°C to 33°C within 24 hours. An effective cooling technique is to reduce daily indoor temperature variations by using the thermal masses of the ceiling and the walls. Materials with a high heat capacity require considerable energy flows to change their temperature, i.e. they dampen temperature variations. Usually, a natural ventilation concept reduces the fresh air infiltration rate to a hygienically appropriate level during the day whereas at night-time cool, nocturnal ambient air is used to cool down the building masses. Effective cooling requires high air exchange rates. Natural ventilation can reduce the cooling load by up to 40% or – if no active cooling is available – reduce the daily peak temperatures by up to 1.5°C. Natural ventilation requires that concrete ceilings and walls be directly exposed to the indoor air. This leads to an architecture that consciously exposes raw materials and the climatization technology instead of hiding it. This may conflict with the acoustic situation, and requires carefully balanced solutions. Figure 4.17(a) sketches the free nocturnal ventilation concept for a low energy office building situated near Stuttgart, Germany. Care has to be taken during the design of such a concept to avoid cost-intensive fire prevention measures, which an open building design can necessitate. The air flow pattern in urban areas, particularly in urban canyons, needs careful assessment, as it defines the real potential for natural ventilation. Appropriate positioning and dimensioning of the openings in order to optimize the resulting air flow is a serious exercise for designers.
- **Controlled ventilation.** An automated ventilation system is usually driven by an electrical fan system, which induces a desired volume of air to flow through a building. The advantage of controlled ventilation over free ventilation is that the quantity and direction of air infiltration are controlled and no unwanted energy exchange with the outside takes place. If incoming as well as outgoing air are sucked in and blown out through a central air pipe, the ventilation system can be combined with an air-to-air heat exchanger to transfer heat or cold from the outgoing to the incoming air. Another option is to use an air-to-ground exchanger to exploit the earth as an energy sink or source for the incoming air (Figure 4.17(b)).

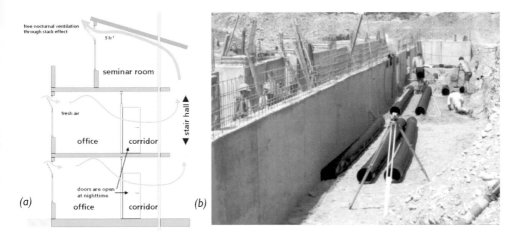

Figure 4.17 (a) Example of free nocturnal ventilation and daylighting concept for a low energy office building. The south-facing offices lie adjacent to a stair hall, which is daylit by a large skylight. All internal walls are glazed above head-level. This horizontal and vertical transparency allows daylight to penetrate deeply into the building, and is a necessary condition for the free nocturnal ventilation concept, which is driven by the 'stack effect': air is forced through slats in the office facades into the circulation area and leaves the building through automatically opened slats in the roof. No electrical fan supports the air movement. The office doors should be opened at night to avoid pressure drops across the door slits. (b) Example of an air-to-ground heat exchanger; the black plastic tubes with a total length of 700 m and a diameter of 25 cm are installed 6 m below the ground. The heat exchanger will support the cooling system for an atrium and adjacent offices with a floor area of about 1000 m².

4.2.2.6 Solar assisted sorptive cooling systems

The number of air-conditioned buildings continues to rise. While at higher latitudes it is mostly commercial buildings that are actively cooled, in the Mediterranean countries residential buildings are equipped with small electrical cooling appliances as well. Most of these cooling systems are electrically powered compression engines with ozone-depleting refrigerants. A search for more sustainable materials and lower electrical loads has led to the development of thermally driven, so-called sorptive cooling systems. As the times of solar availability and high indoor temperatures largely overlap, the prospect of solar-powered cooling systems promises immense energy savings. The working principle of sorptive cooling systems is that water vapour is absorbed by a liquid sorbent or adsorbed by a solid sorbent. Thermal energy is used to desorb the water vapour from the sorbent. In some systems, water is chilled to cool a building via a cooling ceiling or a heat exchanger (closed system); in open systems, the indoor air is chilled directly. Closed cycle systems with solid/liquid sorbents are discussed below:

- **Closed cycle with liquid sorbent.** This sorptive cooling class is commercially available in all system sizes from below 100 kW to several MW. The working pair is water combined with lithium bromide or ammonia. These systems exhibit rising efficiencies with rising driving temperatures. Above 160°C they can outperform compression engines, but at the minimum operating temperatures, around 80°C, they have low efficiencies. They are usually thermally driven by a cogeneration system or a district heating network or even evacuated tube collectors, which can deliver the required temperatures.

- **Closed cycle with solid sorbent.** This system type is commercially available, using a combination of water with a silica gel. While they can operate at driving temperatures as low as 70°C, they are comparatively large, heavy and expensive.

District cooling networks supply participating buildings with chilled water, which is cooled by a central, possibly solar assisted cooling facility. These systems combine financial and environmental benefits, since central cooling systems offer higher annual working hours and performance levels than smaller air conditioning systems. The power peaks of a district cooling system and total power grid do not necessarily overlap, as the chilled water storage tank provides a certain inertia. This leads to a smoothing of the load curve for the central power supply unit.

4.2.3 Retrofits in the existing building stock

The refurbishment of existing buildings provides by far the largest potential for solar building design concepts, for several reasons. The market is immense, because in most EC countries activity and investment in the retrofitting and reuse of buildings surpass those involved in constructing new ones. More than 70% of building-related investments in Western Europe are spent in urban renewal and building refurbishment. In the real estate sector, retrofitting of the post-war stock is considered a major area of activity, as there is a high demand for traditional buildings equipped with modern facilities. The contribution of solar concepts in the expected surge of financial and construction activities could be significant.

Buildings need to be retrofitted to varying degrees and at different time intervals. Motivations for a renovation are diverse: increasing the value of a building, increasing the comfort conditions and facilities for the users, addressing architectural topics, reducing high running costs caused by out-of-date HVAC systems or facade components, adapting the building to new standards or changed routines, and so forth. Obviously, the aim is not generally to find the cheapest solution to maintain the building's proper function, but to find a balanced approach that satisfies various requirements. Although energy is usually not the driving factor, many refurbishment projects offer the opportunity for substantial energy-related improvements. By applying measures for energy conservation in an **integrated renovation approach**, total costs can even be reduced. To achieve this goal, it is important to understand the energy flows in a building *before* renovation, as this allows a cost–benefit analysis of various renovation options. The actual savings potential of various solar renovation concepts has been investigated by the International Energy Agency, Task 20 *Solar Energy in Building Renovation*:[27]

- **Solar collectors.** The roof-mounting of large collector fields during the course of a renovation project can be an attractive option if a central domestic hot water system is available (Figure 4.18(a)). Combination with a district heating system has been described in Section 4.2.1.2. Such projects benefit from the economy of scale. However, insufficiently insulated water pipes and storage tanks often encounter renovation-specific difficulties, which can lead to thermal losses of up to 50%.

- **Glazed balconies.** Balconies with opening glass elements are gaining popularity, as they create additional sun space for the users. Financial synergies are that the glazed cover acts as a protection against facade degradation and reduces thermal bridges. Energetic benefits are sensitive to user behaviour. The balcony must not be actively heated, either indirectly via heat flow from the adjacent room through open doors, or via an electric radiator. Provided this is the case, the balcony can act like a buffer zone between the rooms and the outside. If the balcony is integrated into a controlled ventilation system, it can serve as an air collector. A frequently neglected drawback of glazed balconies is that the additional glazing layer reduces the daylight availability inside the building.

- **Solar walls.** With transparently insulated wall elements, solar gains by far exceed the heat losses during the heating season. Annual heat gains of 50–150 kWh/m^2 can be realized, together with increased thermal comfort if seasonal shading is provided (Figure 4.18(b)).

- **Photovoltaics.** Within the framework of a regular facade renovation, opaque or semi-transparent photovoltaic panels are attractive and innovative facade elements. New solar cells with varying visible transmittances are available. These multi-purpose facade elements provide shading, generate electricity and communicate a positive corporate identity to the outside world, due to the positive image of photovoltaics in society.

(a)   *(b)*

Figure 4.18 Retrofitted building examples. (a) Students' hostel in Dornbirn, Austria. The hostel has been equipped with 112 m^2 of facade-integrated solar collectors with a 2400-litre storage tank. The facade collectors serve as a combined architectural and energy-generating device (architect: Heim & Müller); (b) Villa Tannheim, headquarters of the International Solar Energy society, in Freiburg, Germany. A key architectural element of the retrofitting of the Villa was the facade integration of 52 m^2 of transparent insulation material (architect: Henzel).

- **Daylighting and lean climatization.** There are a range of retrofitting measures available to improve and optimize the energy performance of an existing building, which include redesign of openings, improvement of external opaque elements, microclimate improvements, optimization of the solar control, and improvement of the lighting system by incorporating daylight. Most of these measures resemble those of a lean climatization concept for a new building. In an integrated renovation process these concepts also require a thorough assessment of the surrounding microclimate and the status quo of the building. The planning is challenging, since the boundary conditions are more static than those for a new building.

How can solar retrofitting concepts be financed? The associated costs certainly depend on the specific renovation project. The possibilities and difficulties resemble the situation in new solar buildings: a number of changes can be implemented at little or no extra cost, while further measures require additional investments, which can be financially amortized over a number of years. Capital is rare, and long-term benefits often need to give way to short-term profits. In this context **third-party financing** has great appeal. The concept of this financing scheme is that a contractor organizes and finances an energy renovation concept and guarantees a reduced future energy demand under certain boundary conditions. Depending on the exact conditions, the benefits of reduced operation costs are shared between building owner and contractor until the expenses of the contractor are paid for. Contracting companies are banks or companies that sell large cogeneration energy supply systems. In the latter case, the cogeneration system offers an added benefit that companies can use to convince undecided customers. A recent extension of the contracting concept is **solar contracting**, which is based on the idea that measures with varying financial amortization times are bundled together as a renovation package.

4.3 GOALS FOR R&D

Solar design concepts and components already have an established position in present-day construction practices. Well-insulated windows are a legal requirement in several European countries, the installed collector area is rising exponentially, and daylighting and solar gain control are a central design aspect in new office buildings with large architectural glazings.

Further progress requires additional R&D activities on components and concepts as well as educational colloquia. A number of building components, which will be specified below, need further development for successful commercialization. New testing and durability assessment procedures have to be developed for these components to accelerate their entrance into the market. The integration of these products into existing energy supply systems is a necessary condition for a wider market penetration. At the same time, simulation tools need to be developed that can predict both the influence of the user on the final energy demand of a building, and the interaction of the different system components so that efficient control algorithms can be developed. More research on the sociological and financial aspects of an integrated design process is necessary, to identify structural barriers and sources of conflict between the parties involved. The resulting findings should be integrated into the educational colloquia of architects and building engineers. Finally, extended monitoring of exemplary buildings is needed to identify weaknesses in building energy concepts for further research activities.

4.3.1 Components

As explained above, lean cooling concepts are in need of advanced solar gain control technologies and strategies. The following components are on the verge of entering the market. Apart from technical shortcomings such as long-term stability, the architectural integration of switchable glazings also deserves attention.

- **Angle selective coatings.** The basic idea of an angle selective coating is to create a window with a high reflectivity of incident light at high solar altitudes, and yet a high transmittance of light incident at near-horizon angles. In this way, visual contact with the outside can be provided while excessive heat gains and glare from direct sunlight or sky-scattered light are reduced. In technical terms, a microscopic structure in the form of inclined columns is added on a glazing surface, and these columns act for a portion of the incoming sunlight as small venetian blinds tilted downwards.
- **Thermotropic and thermochromic coatings.** The transmittance of thermotropic and thermochromic coatings is temperature-dependent, and exhibits a distinct fall above a certain threshold temperature. A thermochromic coating consists of a transition metal layer that switches from a semiconductor or dielectric state to a metal state above a fixed temperature. A thermotropic layer consists of a mixture of either two polymers or a polymer with water. The two utilized components have different refractive indices. Above the threshold temperature they segregate, so that the usually transparent material becomes opaque. In combination with a low-e coating, a visible transmittance range of a polymer blend from 73% at 30°C to 31% at 50°C has been realized.[24] Figure 4.19 shows the switching process of a thermotropic glazing whose architectural suitability in an office facade has been investigated. Although demonstrated here as a window glazing system, the main applications of thermotropics are for units with no need for individual control, e.g. skylights. Apart from further research activities on the long-term stability of thermotropic glazing units, it will be necessary to investigate different architectural applications for the system, to assess both the energy saving potential and user acceptance of a thermotropic facade element.
- **Gasochromic glazings.** A gasochromic glazing unit also features a variable solar and visible transmittance. It consists of two glazings with a coating on one of the two inner surfaces. The coating is the optically active layer and consists of a WO_3 film that turns blue when exposed to low concentrations of hydrogen. Exposed to oxygen, the colour bleaches again. Under a reference AM 1.5 solar spectrum, a solar transmittance range from 11% to 74% has been realized. In a prototype sample,

Figure 4.19 Picture study of a scaled prototype facade with thermotropic glazing: as the ambient temperature rises, the thermotropic layer becomes opaque.

the colouring took some 20 seconds, while the bleaching required about 1 minute.[25] The rate of coloration, as well as the transmittance of the resulting glazing unit, can be changed. As the switching signal is external, the users can influence the transmittance of the glazing according to their needs, and a high visual comfort is expected at all times.

Present research activities aim to render the coloured state of a gasochromic layer grey instead of blue by doping the WO_3 layer with other transition metals. Other efforts are being undertaken with the aim of developing a closed control system that provides the necessary gases by using an integrated electrolyser.

- **Photochromic glazing.** Photochromism is induced by ultraviolet radiation, which (reversibly) changes the energy state of a photochromic material – usually a metal halide or a plastic. Photochromic coatings are a well-known product in spectacles, but they are not presently used in buildings. They are too cost-intensive, and the effect on the g-value is limited as the solar radiation is absorbed in the glazing and not reflected on the surface. The UV and temperature stability of photochromic plastics remain in the research phase.
- **Accelerated testing and ageing procedures.** Innovative building components must exhibit lifetimes of several decades, often under rough climatic conditions. New methods are needed to test the long-term stability of material surfaces and components to ensure that they are fit for the market.
- **Multi-functional surfaces.** Apart from low emission coatings (Section 4.2.1.1), other surface technologies are being developed that modify the optical or adhesive properties of the treated surface, which becomes minimally reflective or self-

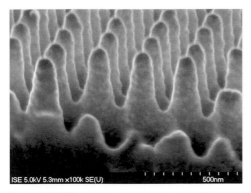

Figure 4.20 Electron microscopy picture of a
self-cleansing structure.

Figure 4.21 Visualization of an anti-reflective
coating: solar absorber plate with a glass cover.
The lower part of the glass cover has been
finished with an anti-reflective coating.

cleansing. Figure 4.20 shows an electron microscopy picture of a self-cleansing
surface while Figure 4.21 visualizes the effect of an anti-reflective coating. Technical
issues like stability, deposition techniques for large area surfaces and low system
costs dominate future research activities on these technologies.

- **Multi-functional facades.** Solar gain control is only one performance aspect of a
facade among others. Modern facade elements aim to provide demand-tailored
functions. They serve as daylight elements, control solar gains, and feature ventilation
slits and sometimes photovoltaics.

  New construction methods are necessary to assemble the required elements in a
single facade that meets all applicable sound, fire and mechanical specifications at
a reasonable cost. The interaction of a multi-functional facade with the daylight
and climatization concept of the whole building is complex, and there is an urgent
need for suitable evaluation methods to characterize the overall performance of the
facade as well as to identify suitable control algorithms.

- **Collector technology.** Collectors are used for various purposes ranging from low
to high temperature applications. To satisfy these diverse requirements, further
material research on both the transparent cover and the absorber is necessary. The
long-term stability and environmental impact of these new materials is an urgent
research issue. The energy efficiency of circulation pumps needs to be increased.
Control algorithms for large solar applications have to be developed, and stagnation
effects in large collector fields, which can lead to permanent system damage, need
to be understood so that they can be avoided in the future.

Another product development aims to combine photovoltaic and collector systems in homogeneous-looking hybrid systems that require less space and can be more easily integrated into a roof or facade design. More aesthetically pleasing collectors will lead to a higher acceptance of collectors by building owners and architects.

- **Long-term storage.** Cheap seasonal storage systems for hot water are required to increase the solar contribution to the heating demand of buildings. At present, the heat losses during transport and storage of hot water can make up 30% of the total primary energy use of heating systems. An improved system layout using better insulation techniques and shorter pipe lengths can reduce energy losses. A new technical development is the investigation of the seasonal storage of solar energy, via the adsorption of water vapour on adsorbents like silica gel.[28] This exciting technology will allow the seasonal storage of low temperature heat at energy densities that are already four to five times higher than that of water. These laboratory concepts need to be tested in demonstration projects.
- **Phase change materials.** These are innovative plaster materials whose heat capacity exhibits a distinct increase at around 24–28°C. This physical property allows them to act as a barrier towards higher indoor temperatures. A promising application of these innovative materials is to integrate them into a climatization concept, as they can reduce daily indoor temperature variations.
- **Low-power heating systems.** To meet the remaining energy demand of low energy houses, suitable cogeneration technologies need to be examined. Examples of such systems are small heat pumps, and fuel cells for stationary applications. These components have to be carefully integrated into existing energy supply systems.
- **Solar combi-systems.** The inherent complexity of solar combi-systems – outlined in Section 4.2.1.3 – has led to a number of widely different, commercially available system designs. These design solutions result mainly from field experiences and have not been carefully optimized, although some of them have the potential to be sold across Europe. For the market expansion of these systems to take place, standardized test procedures need to be developed, to assess the energy performance of different solar combi-systems under varying climatic boundary conditions and applications.
- **Solar assisted cooling.** Open cycle cooling systems with a solid sorbent have the potential to become a financially viable alternative to compression machines for creating climatization. The operating principle is that incoming air is dried via direct contact with the solid sorbent (Figure 4.22). The dry air is afterwards chilled by adiabatic cooling, i.e. as the relative humidity of the air increases, its temperature falls. The whole process is called **desiccant cooling**. Systems with liquid sorbents presently exist only at the laboratory stage, while adsorbing systems strive for a wider market penetration. The environmental charm of these systems is that operating temperatures around 55°C already yield satisfying efficiency levels. This low temperature range can be provided by flat plate collectors, which could be used for heating in the winter and cooling in the summer.

Figure 4.22 Highly efficient heat exchanger matrix of copper lamellae with an integrated solid sorbent (silica gel) for applications in adsorption heat pumps or coolers.

4.3.2 Planning tools

- **Lean cooling systems:** To develop a lean cooling concept for a building, reliable simulation tools are necessary to model the underlying aerodynamics. Present problem areas are natural ventilation systems and daylit atria. Further research is needed to identify suitable boundary conditions for typical room situations and urban settings.

- **Modelling comfort.** Whereas thermal comfort has been widely investigated, little scientifically sound information is available on the perception of visual comfort. This lack of information leads to considerable uncertainties in simulations of the energetic performance of daylight elements, shading devices, artificial lighting and other user-operated components. Field studies are needed, and based on these, further information about visual comfort can be collected and transferred into simulation models.

- **Integrated performance tools.** As described above, building simulation tools are becoming increasingly powerful as they consider the multiple physical phenomena that influence the energy flows in a building. Innovative tools are needed, able to simulate advanced control strategies of integrated energy concepts to allow a reliable dimensioning of the system components for a given building. Other aspects that should be covered by future simulation tools are life-cycle and cost–performance analysis. Future research should also cover the intuitive visualization of simulation results, to facilitate communication between engineers and architects in the design process.

4.3.3 Innovative educational colloquia

An integrated design process requires ways of communication that are fundamentally different from the regular design process. In the latter, initial communication is mainly between the building owner and the architect. Based on the building design, the architect formulates requirements for the climatization system and the building envelope, which are then provided independently by the various responsible engineers. No direct communication between the various parties is planned for. The intended exchange of ideas in an integrated design process can easily create friction, as the areas of competence and responsibility are not as clearly defined. To overcome fears, the educational colloquia of the involved parties need to be adapted to the changing demands.

- Architects should understand that their role in the design process is as a mediator between the needs of the building owner and the technical necessities formulated by the building engineers. Therefore a closer understanding between architect and engineers is necessary.
- Basic simulation tools can help to heighten architects' awareness of solar building design principles.
- Large-scale solar applications need specially educated craftsmen who can properly and efficiently install these technologies.

4.3.4 Monitoring and evaluation

Intensified field studies are necessary to test the effectiveness of single components and whole design concepts in 'real world' projects. A careful analysis of the results can help to identify the weaknesses of existing concepts and components. Further efforts must be made to find a common terminology for the presentation of building data, as national codes vary significantly.

- **Commissioning/prehandover testing.** To ensure that the energy concept of a building works properly, it is crucial that the system is thoroughly tested and adjusted according to the needs and habits of the users. Such system adjustments are usually only provided with minor efforts on behalf of the HVAC engineer, as they constitute an additional workload. Nevertheless, in order to ensure that innovative building concepts function fully, it is necessary to develop simple testing schemes that can be carried out within a suitable timeframe.

4.4 ROADS TO A STRONGER MARKET

The building sector is constantly subject to changes. The nature of these changes is part of a larger public discussion about how we live and work together and share our resources. Across Europe, the individual nations follow different building policies, but it is desirable that diverse national solutions fit into a wider *trans-European* context.

The sections above have demonstrated that buildings and the layout of urban centres are a key element in the energy concept of our society. Even today, more than 10% of all energy in the building sector required for space heating and domestic hot water is provided by solar energy. Every household influences whether building-related carbon emissions rise or fall. A catalogue of innovative technologies and concepts is available today, and their implementation would lead to more sustainable building practices. Some ideas have turned into commercial products, others are on the verge of commercialization, and still others require further research. A new kind of architecture is emerging that aims to combine energy efficiency and solar energy supply measures with high-quality architectural design and excellent comfort conditions (Figure 4.23). Further examples of this building species will convince a wider public of the various benefits of solar building design. Even though the greatest energy saving potential lies in the existing building stock, more sustainable building practices for new buildings are also needed, since 1% of the building stock is newly constructed every year.

It is probable that the positive trend in the residential single-family building sector will continue, as the benefits for the individual building owner are already substantial. Integrated solar heating systems are financially viable and technically mature, and the benefits of healthy housing with reduced future operation costs are a recognized savings plan for the future. Investors in apartment buildings will adapt to this trend once the facts that solar building design concepts work and that a market for solar housing exists are more widely disseminated. Such large-scale applications will replace fossil fuels for space heating and domestic hot water. To meet the rising electricity demand of commercial buildings, solutions that combine excellent comfort conditions with reduced energy use are necessary – lean climatization concepts and increased daylighting are candidates for this purpose. The construction costs for a lean commercial building already lie within the price range of conventional commercial buildings, and the extra planning costs will fall once an integrated design process has become more common. Motivated municipalities or regional governments have various options at their disposal to foster sustainable building practices in the areas under their authority.

(a)

(b)

Figure 4.23 (a) Athmer office building, Germany.This building won the architectural award for low energy buildings with high-quality design.A key architectural element of the building is an atrium, which creates a bright and friendly entrance area (architects: Banz + Riecks).(© K Ortmeyer) (b) AVAX SA headquarters, Greece.This building was part of the THERMIE Energy Comfort 2000 programme, which was partly funded by the European Commission.The main aim of the design was to apply bioclimatic design features in order to cover the energy demands of the building and create a comfortable environment for the users (architect Tombazis).

Such development requires a clear and well communicated plan as to how the region should develop. Balanced spatial planning includes mixed residential, industrial and recreational areas. District cooling and heating are natural options for such regions, and well organized public transport can keep total emission levels per capita low. Finally, setting the right example by making public buildings lean and solar is the most convincing publicity for low-energy solar buildings.

A consistent, Europe-wide action plan would accelerate the take-up of solar building concepts and close the development gap between the EU members. Historically, the setting of permanent and favourable legislative boundary conditions has triggered a surge of private activities. A prominent example is the explosive development of the wind industry in Germany following a law that guarantees a minimum price for renewable energy. Four policy actions, outlined below, would contribute to more sustainable building practices:

- benchmarking
- quality control
- further research
- retrofitting.

4.4.1 Benchmarking: a call for new building codes

The significance of a building's electrical energy demand is not, at present, widely recognized. A major reason for this lack of recognition is that current electrical energy prices do not reflect environmental and economic realities; external costs of fossil energies are not paid for by the benefiting company or population group, but they constitute a burden to society as a whole. Falling electricity prices in the European market underline this tendency.

Recent developments in EU member states with strict building codes have shown that legal requirements can succeed in initiating a positive trend in thermal energy efficiency. The same policy measure could be employed if electrical energy benchmarks were to join thermal regulations in extended building codes. The ambitious electrical benchmarks from the Swiss SIA are notable guidelines for specifying thermal and electrical energy use by buildings (Figure 4.12). Benchmarks define boundary conditions comparable to a cost budget, without impairing the creative freedom of the architect. The final goal, not the means, would be regulated.

An alternative to introducing building codes at a single building level is the introduction of benchmarks at the district level. This would relieve the strain on the individual building owner, and provide municipalities with greater freedom in laying out their energy supply systems.

4.4.2 Quality control

A building code invents a rule but does not define reality. Therefore, a certain degree of quality control is required, to ensure whether or not energy goals have been achieved.

Present public funding schemes pay for installed equipment and not for avoided emissions. Some buildings receive high publicity without fulfilling the expectations they were invested with. A possible control mechanism is the implementation of electrical counters at crucial positions in a building, which would measure the energy demand for heating, cooling and lighting, excluding user-specific demands. This proceeding would:

- allow the building owner to check whether actual energy demands reflect predictions made during the design phase
- provide building owners and architects with an idea of how sustainable their buildings are in comparison with others
- allow a scientifically sound analysis of which concepts work or not and why.

4.4.3 Further research

Many of the above-mentioned building concepts and components are already financially viable or convincing in the laboratory. However, building research must not be limited to refining what already works. New ideas must be considered and tested before they reach the market stage. This task requires funding for basic applied research that is independent of external industrial pressure, milestones and short-term goals. Co-operation and dialogue between industry and public research centres must be maintained, but dependence of the latter on the former will prove fatal in the long term. Research must be funded in concepts that are still at an early phase, so that they can enrich the range of energy saving measures at a later stage.

4.4.4 Retrofitting: the road not taken

'One can make mistakes but must not build them' (J W von Goethe). Most buildings outlive the times and shortcomings of their creator. Changes in the building sector take considerable time to show their effect. Various methodologies for integrating energy saving measures during refurbishment have been proposed and tested, and the accompanying energy saving costs are often extremely low. It is important to develop financing models for these applications and enact legislation to promote them. This task of overcoming the non-technical barriers to sustainable building development is one of the pressing challenges of our time.

REFERENCES

1   ESAP sa, *Energy in Europe: 1998 – Annual Energy Review*, report for the European Communities, 1998.
2   M Santamouris and A Argiriou, 'Renewable energies and energy conservation technologies for buildings in southern Europe', *International Journal of Solar Energy*, Vol. 15, 1994, pp. 69–79.
3   J Wienold, O Lewis, P Lund, G Mihalakakou, M Santamouris, M Thyholt, L Vandaele, K Voss, M Wangusi, and S Burton, *Contribution and Potential of Passive Solar Gains in Residential Buildings in the European Union*. Contract No. STR-1742-98-EU, January 2000.

4   J-O Dalenbäck, *Solar Thermal Systems*. Conference Proceedings of EUROSUN 2000, Copenhagen, Denmark, 19–23 June 2000.

5   E O'Cofaigh, E Fitzgerald, R Alcock, A McNicholl, J O Lewis, V Peltonen and A Marucco, *A Green Vitruvius: Principles and Practice of Sustainable Architectural Design*. James & James, London, 1999.

6   A M Hestnes, 'Building integration of solar energy systems', to be published in *Solar Energy*, 2000.

7   TEMIS (Total Emission Model of Integrated Systems), Version 3.x, Ökoinstitut, Darmstadt, Germany, 1999.

8   M N Fisch, M Guigas and J O Dalenbäck, 'A review of large-scale solar heating systems in Europe', *Solar Energy*, Vol. 63, No. 6, 1998, pp. 355–366.

9   G Stryi-Hipp, 'Solarwärmenutzung in Europa', *Erneuerbare Energien*, Vol. 4, 1998, pp. 4–9.

10 DLR 1998 (German national aerospace research centre), http://www.dlr.de

11 A Nilson, R Uppström and C Hjalmarsson, *Energy Efficiency in Office Buildings: Lessons from Swedish Projects*, Swedish Council for Building Research, 1997.

12 *Schweizer Energiefachbuch, Bundesamt für Statistik BFS: ein beispielhafter Bundesbau*, Künzler–Bachmann, St Gallen, 1999, pp. 47–49.

13 N Baker, 'Environmental comfort: optimization or opportunity', *Proceedings of Low Energy Building Conference '99*, Hamburg, 17–18 September 1999.

14 *Energie im Hochbau SIA 380/1-4*, Guideline for energy use in commercial buildings. Association of Swiss Engineers and Architects (SIA), Zurich, 1995.

15 C A Redlich, J S Sparer and M R Cullen, 'Sick-building syndrome', *The Lancet*, Vol. 349, 1997, pp. 1013–1016.

16 P Lohr, *Das Kölner Low Energy Office: ein Bürogebäude im Niedrigenergie-Standard*. Annual report Implusprogramm Hessen, Germany, 1998.

17 Esbensen Consulting Engineers, *Energy Consumption and Cost Effectiveness of EC2000 Buildings*, report for the European Commission, 1998.

18 BKI, *Baukostenindex Teil 1*. Baukosteninformationszentrum, Stuttgart, 1999.

19 J Clarke *et al.*, 'Prospects for truly integrated building performance simulation', *Proceedings Sixth International IBPSA Conference (BS '99)*, Kyoto, Japan, September 1999, Vol. 3, pp. 1147–1154.

20 W A Beckmann, *TRNSYS: A Transient System Simulation Program*, Solar Energy Laboratory, University of Wisconsin-Madison, USA, 1994.

21 D B Crawley *et al.*, 'ENERGYPLUS: a new-generation building energy simulation program', *Proceedings of Conference on Renewable and Advanced Energy Systems for the 21st Century*, April 1999, Maui, Hawaii, USA, 1999.

22 G Ward, *The RADIANCE 2.4 Synthetic Imaging System*, University of California at Berkeley, USA, 1994.

23 J Ludwig, H Quin and B Spalding, *The PHOENICS Reference Manual: Revision 08. Software Version 1.51*, CHAM report number TR/200, 1989.

24 H Wilson, 'Solar control coatings for windows', *Proceedings of EUROMAT 99*, 27–30 September 1999, Munich, Germany.

25 H Wilson, A Gombert, A Georg and P Nitz, P, 'Switchable glazing with a large dynamic range in total solar energy transmittance (g-value)', *Conference Proceedings World Renewable Energy Congress – VI (WREC 2000)*, 1–7 July 2000, Brighton, England.

26 T Deinlein, Pilkington Flabeg GmbH, 90766 Fürth, Germany, Fax: +49 911 9974450, 2000.

27 K Voss, 'Solar energy in building renovation: results and experiences of international demonstration buildings', accepted for publication in *Energy and Buildings*, 2000.

28 W Mittelbach and H-M Henning, 'Seasonal storage using adsorption processes', *IEA Workshop on Advanced Solar Thermal Systems*, September 1997, Helsinki, Finland.

5. SOLAR THERMAL POWER PLANTS

*M Becker, W Meinecke, M Geyer, F Trieb (DLR), M Blanco,
M Romero (CIEMAT), A Ferrière (CNRS)*

5.1 INTRODUCTION, POTENTIAL AND STRATEGIC SUMMARY

In solar thermal power plants, the incoming radiation is tracked by large mirror fields, which concentrate the energy towards the absorbers. These in turn receive the concentrated radiation and transfer it thermally to the working medium. The heated fluid operates as in conventional power stations, either directly, if steam or air is used as the medium, or indirectly, through a heat-exchanging steam generator on the turbine unit, which then drives the generator.

Various concentrating technologies have been developed, or are currently under development, for commercial applications. They are designed to make the high solar flux with a high energetic value that originates from processes occurring at the sun's surface (at black-body-equivalent temperatures of around 5800 K) usable in technical processes and commercial applications. Such solar thermal concentrating systems will undoubtedly make a significant contribution to efficient, economical and clean energy supply in the next decade.

This chapter deals with three different technologies for solar thermal power plants that make use of concentrating solar energy systems (see Figure 5.1):

- parabolic troughs
- central receivers (towers)
- parabolic dishes.

Solar thermal power plants represent one of the most cost-efficient options for renewable bulk power production. After successful technology demonstrations and some commercial success (354 MW_e of parabolic trough systems) a decade ago, the strategic approach for widespread application of these technologies may be summarized as follows:

- Build confidence by realizing a number of pilot applications on the basis of conservative designs and proven technologies. Hybrid solar/fossil fuel plants will

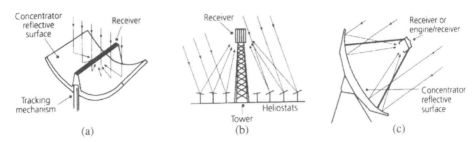

Figure 5.1 Schematic diagrams of the three main concepts for concentrating solar thermal electricity generation: (a) parabolic trough, (b) central receiver (tower), (c) parabolic dish. (Source: DLR-Köln)

at this stage reduce the financial and operational risks for potential investors.

- Reduce cost by developing and demonstrating improved components, subsystems and processes. Increasing production quantities due to the growing numbers of installations will also help to exploit economies of scale.
- Increase the solar share by suitable process design and storage integration. The ultimate goal in the future will be an economic solar-only operation.

Apart from component and subsystem improvements, where significant synergies exist (e.g. reflector materials, tracking control), the competing technologies have different starting points and prospects, and face different challenges. These are summarized below.

In view of the promises and risks associated with the different technologies, none should be neglected at the present stage of development. Parallel development will not only exploit the synergies mentioned above, but will also stimulate competition, until market forces indicate the most promising routes.

In the European Commission's 1997 White Paper on a community strategy and action plan for renewable energies, it was proposed that at least 1 GW_e of these systems should be implemented in Europe by the year 2010. This objective can be achieved by establishing 25–30 commercial solar thermal power plants of 30–50 MW_e each, distributed throughout the south of Europe.

These solar thermal technologies comply with the prime objectives and key research, technology and demonstration actions of the Fifth Framework Programme of the European Commission. This is because:

- their development will enhance the deployment of solar energy systems for bulk electricity production and the reduction of energy from fossil fuel sources, with a consequent reduction of impact on the environment, particularly with respect to their high potential for contributing to the reduction of CO_2 emissions
- they reduce the generating costs of solar power plants, and thus help to ensure durable and reliable energy services at affordable costs in the medium to long term

- they will put European industry in a privileged technological position, thus opening possibilities for industrial growth, not only for the internal market of southern European countries, but also for the export of equipment and services in the field of solar plant installations; in most cases the solar-specific technology required is also accessible to local industries, and the opportunity to create links with developing countries and their markets is self-evident
- they have the potential to contribute to the social objectives of the European Union relating to quality of life, health, safety (including working conditions) and job creation; the establishment of such plants in undeveloped areas in southern Europe can create new opportunities for industrial fabrication, assembling, operation and maintenance.

Solar thermal power plants are important candidates for providing a major share of the clean and renewable energy needed in the future. This is because:

- they are among the most cost-effective renewable power technologies, promising to become competitive with fossil fuel plants within the next decade
- they represent a proven and well-demonstrated technology; since 1985 nine parabolic trough-type solar thermal power plants in California have fed over 8 billion kWh of electricity into the Southern Californian grid, demonstrating the soundness of the concept
- they are now ready for intensified market penetration; accelerated grid-connected applications will lead to further innovation and cost reduction.

However, no new commercial solar thermal power plants have been built since the last two 80 MW$_e$ parabolic trough plants (SEGS VIII and IX) were connected to the Southern California grid in 1991 and 1992 respectively. The main reasons for this are as follows:

- the financial uncertainties caused by delayed renewal of favourable tax provisions for solar systems in California
- the financial problems and subsequent bankruptcy of the US/Israeli LUZ group, the first commercial developer of private solar power projects
- the rapid drop in fossil energy prices and subsequently stable energy prices at low levels worldwide
- the large unit capacities of solar thermal power stations required to meet competitive conditions for the generation of bulk electricity, resulting in financing constraints due to the inherently large share of capital costs for solar installations
- the rapidly decreasing depreciation times of capital investment for power plants due to the deregulation of the electricity market and the shift to private investor ownership of new plant projects worldwide
- dropping specific prices and enhanced efficiencies of installed conventional power plants, particularly combined-cycle power plants

- the lack of favourable financial and political environments for new solar thermal power plant project development in sunbelt countries.

Fortunately, the chances for solar thermal projects in the next decade have increased significantly in the meantime, because:

- interest rates and capital costs have fallen drastically in Europe with the introduction of the euro, to the benefit of capital-intensive projects such as solar thermal power plants
- EU policy firmly supports an 8% reduction in gas emissions by doubling the share of renewable energies in the EU energy balance from 6% to 12% by the year 2010
- recent World Bank predictions suggest that there will be solar thermal cost reductions below US$0.06/kWh after 2010, convincing the Global Environment Facility (GEF) to support barrier removal and market introduction of solar thermal power in developing countries
- India, Egypt, Morocco and Mexico have now applied within the GEF Operational Programme for about US$50 million of GEF grants for each project to cover their incremental costs for solar thermal power projects.

With positive experiences in the construction and operation of the first European demonstration power plant projects – the 50 MW_e THESEUS on the island of Crete in Greece, and the 10 MW_e Planta Solar (PS 10) and the 12.5 MW_e SOLAR TRES in southern Spain – other projects are expected to follow. Until the year 2015, the market potential for solar thermal power plants is estimated to be at least 7 GW_e in southern Europe, representing a potential for CO_2 reduction of up to 12 million tonnes per year.

These projects represent a cost reduction potential of 20% compared with the last plant built, the 80 MW_e SEGS IX plant in California. Projected electricity production costs can then come down to €0.14/kWh (in pure solar mode, without any grant). However, electricity costs of solar power plants operating in the solar/fossil hybrid mode (as encouraged by the EU Fifth Framework Programme) could fall to as low as €0.08 /kWh in the short term. Research and development programmes in Europe and in the USA are aimed at reducing electricity costs further in the long term.

The Mediterranean member states of the European Union will surely be counting on solar thermal power plants as an excellent option for achieving the goals in their national policies. Within this framework, it is relevant to mention the recent approval of a promotion plan in Spain that foresees the installation of at least 200 MW_e in the country by 2010, with an annual power production of 413 GWh. It is expected that the new Spanish legislation on renewable electricity generation will allow solar thermal power plants to earn an additional premium on top of the conventional electricity price, which will undoubtedly favour the short-term construction of commercial solar thermal power plants in Spain.

5.2 ACHIEVEMENTS, PRESENT SITUATION, MAIN BARRIERS AND VISION FOR THE WAY FORWARD

5.2.1 Parabolic trough systems

Trough systems use linear concentrators of parabolic shape with highly reflective surfaces, which can be turned in angular movements towards the sun and concentrate the radiation onto a long-line receiving absorber tube (see Figure 5.1). The absorbed solar energy is transferred by a working fluid, which is then piped to a conventional power conversion system.

The power conversion systems are based on the conventional Rankine cycle/steam turbine generator or on the combined cycle (gas turbine with bottoming steam turbine). Trough power plants are highly modular, and are already being applied up to 80 MW$_e$ unit capacity using a thermal oil heat transfer system. Total power plant capacities of above 100 MW$_e$ are actually projected for solar/fossil 'hybrid', integrated solar combined-cycle systems (ISCCS), with equivalent solar field capacities of 30–40 MW$_e$, to help introduce the systems in the energy markets of sunny countries in the near to medium term (e.g. in the countries with GEF support mentioned above).

5.2.1.1 Energy service sector and role of technology

At the moment, parabolic trough power plants represent the most mature solar power plants, with 354 MW$_e$ of commercial solar electric generating systems (SEGS) using parabolic trough power plants connected to the Southern California grid since the 1980s (Figure 5.2). These plants have unit capacities of 14 MW$_e$ (SEGS I), 30 MW$_e$ (SEGS II to VII) and 80 MW$_e$ (SEGS VIII and IX). With more than 2 million m^2 of total glass mirror area of current trough technology, they have generated over 8 TWh of electricity

Figure 5.2 Parabolic trough field arrays of five SEGS plants at Kramer Junction, CA./USA. (Source: Pilkington Solar International, Germany)

since 1985. The 30 MW SEGS plants of Kramer Junction, with an annual insolation of over 2700 kWh/m², have verified generating costs of US$0.15/kWh during high-priced daytime slots (mainly for peak-load demand from the use of air conditioning), with the allowance to generate up to 25% of the annual output by supplementary natural gas firing. (The equivalent pure solar costs would be US$0.20/kWh.) The 80 MW SEGS plants of Harper Lake, with the same annual insolation, have verified generating costs of US$0.12/kWh, also with an allowance to generate up to 25% of the annual output by supplementary natural gas firing. (The equivalent pure solar costs would be US$0.16/kWh.) They have gained record values of an annual plant efficiency of 14% and a daily solar-to-electric efficiency near 20%, as well as peak efficiencies up to 21.5%. The annual plant availability exceeded 98%, and the collector field availability more than 99%. The performance of the plants has been improved continually during the operation time. During the last five years, the five plants at the Kramer Junction site (SEGS III to VII) have achieved a 30% reduction in operation and maintenance costs. In addition, key trough-component manufacturing companies in Europe and associated partners have made considerable advances.

Parabolic trough plants may be designed as quasi-solar-only plants for peak load (as the SEGS plants are, with about 2000 equivalent solar full-load hours per year, and 25% supplementary fossil firing) or, in the future, on an annual average of up to 100% solar share, by applying thermal energy storage systems. Plant concepts favourable for the market introduction are hybrid ISCCS plants for mid-load or base-load operation, with solar shares between 10% and 50%.

5.2.1.2 Technology shortages
Parabolic trough systems have the following technology limitations:

*   The upper process temperature is currently limited by the heat transfer thermal oil to 400°C.
*   The heat transfer thermal oil adds extra costs for investment, operation and maintenance.
*   Depending on national regulations, environmental constraints from ground pollution by spillage of thermal oil could occur.
*   Some absorber tubes are still subject to early degradation, because there is the risk of breakage of the absorber envelope glass tubes, with the loss of vacuum insulation and degradation of the absorber tube selective coating.
*   High winds may break the mirror reflectors at field corners.
*   Low-cost and efficient energy storage systems have not been demonstrated as yet.
*   The direct steam generation trough technology is still in a developmental stage.

5.2.1.3 Current projects
While the European Commission currently supports pure solar plants, the World Bank/ GEF focuses on the integration of parabolic troughs into combined-cycle plants (ISCCS) in sunny developing countries. Various European and US project developers' commercial

power plant development projects, with unit outputs of 50–310 MW$_e$ and large solar fields of parabolic trough collectors, are currently being promoted or are in a progressive planning stage, with grants from the World Bank/GEF and various other worldwide co-funds. Specifically, these are located in:

- **Greece.** 50 MW$_e$ THESEUS solar thermal power plant on Crete; promoted by German and Greek companies; solar field of approximately 300,000 m^2, 112 GWh of pure solar electricity per year (EU co-funded under FP 5).
- **Spain.** 32 MW ANDASOL parabolic trough plant in the province of Almeria, promoted by prominent Spanish and German firms. The Andasol plant will for the first time use the EuroTrough technology in its 235,000 m^2 solar field, and will incorporate a 16,000 m^2 subfield with direct solar steam generation of the DISS technology (EU funded).
- **Egypt.** 135 MW$_e$ natural-gas-fired ISCCS plant in Kuraymat on the Nile river; 30 MW$_e$ equivalent solar capacity; promoted by industrial groups; with allocated US$40–50 million GEF grant.
- **Morocco.** 150 MW$_e$ natural-gas-fired ISCCS plant project; 30–50 MW$_e$ equivalent solar capacity; promoted by industrial groups; with allocated US$40–50 million GEF grant.
- **India.** 140 MW$_e$ naphtha-fired ISCCS plant in Mathania/Rajasthan; 35 MW$_e$ equivalent solar capacity; promoted by industrial groups; with allocated US$49 million GEF grant and US$100 million loan from the German KfW Bank.
- **Iran.** Feasibility study for the implementation of a 100 MW natural-gas-fired combined-cycle plant with a 200,000–400,000 m^2 parabolic trough field in the desert of Yazd, contracted with its own national funds.
- **Mexico.** 310 MW$_e$ natural-gas-fired ISCCS plant in the Northern Mexican desert; 40 MW$_e$ equivalent solar capacity; promoted by industrial groups; with allocated US$40–50 million GEF grant.

The German/Spanish R&D project DISS generates direct steam using LS-3 trough collectors at the PSA in southern Spain, with co-funding from the European Commission (schedule 1996–2001; see Figure 5.3). This new concept is expected to produce a cost reduction over the SEGS plants of 20–30%. Future solar-only solar thermal power plants are planned, with potential use of solar energy storage systems in order to enlarge the solar capacity or solar share and ultimately to minimize CO_2 emissions. The EuroTrough R&D project of a European group under Spanish industrial leadership has been in progress with EU co-funds since 1998, with the goal of reducing the costs of an advanced European trough collector on the basis of the proven LS-3 collector type.

5.2.1.4 Competitors, cost situation and predictions of cost range
Competitors are the current conventional grid-connected fossil-fuel-fired power plants, and particularly the modern natural-gas-fired combined-cycle plants in mid-load or base-load operation mode.

Figure 5.3 Direct solar steam generation test facility using LS-3 trough collectors at the Plataforma Solar de Almería. (Source: CIEMAT-PSA, Spain)

Installed plant capital costs of the Californian SEGS Rankine cycle trough systems with on-peak power operation fell from US\$4000/kW$_e$ to under US\$3000/kW$_e$ between 1984 and 1991, mainly due to scaling-up from 30 MW$_e$ to 80 MW$_e$ units and series effects. The 50 MW$_e$ THESEUS plant will already meet the near- to mid-term cost targets set out in the EU Fifth Framework Programme for solar power systems with €2500/kW$_e$ installed. Projected electricity production costs for a new 50 MW parabolic trough plant at a Southern European site, with an annual insolation of 2400 kWh/m² (as on the island of Crete) are then €0.14/kWh (in pure solar mode without any grant), or €0.18/kWh at a site with 2000 kWh/m² per year, as in Southern Spain. However, in hybrid mode with up to 49% fossil-based power production, the electricity costs could drop to as low as €0.08/kWh.

Installed trough field costs have dropped to €210/m² of current or short-term installed collector technology (based on the LS-3 type). They are expected to fall below €200/m² for enhanced collectors and larger production rates in the medium term, and to €110–130/m² for high production rates in the long term: a 15% discount on the US/European price level of solar installations may be projected for developing countries, owing to lower labour costs.

As analysed by the World Bank, installed plant capital costs of short-term trough plants are expected to be in the range €2440–3500/kW$_e$ for 30–200 MW$_e$ Rankine cycle (SEGS type) plants, and about €1080/kW$_e$ for 130 MW$_e$ hybrid ISCCS plants with 30 MW$_e$ equivalent solar capacity for the US/European construction price scenario. The projected total plant electricity costs range from €0.07 to €0.10 per kWh for SEGS-type plants, and are less than €0.07/kWh for ISCCS plants, which are not however assessed to have electricity costs lower than conventional gas-fired combined-cycle plants.

The expected drop in installed capital costs for grid-connected ISCCS trough plants will result in electricity costs of €0.06/kWh in the medium term, and €0.05/kWh in the long term (200 MW$_e$ Rankine cycle plant with and without storage). There is promising long-term potential for Rankine cycle trough plants to compete with conventional peaking Rankine cycle plants (coal fired or oil fired) at good solar sites. The long-term application of direct steam generation trough technology has even higher potential owing to its promising cost reduction.

5.2.2 Central receiver systems

Central receiver systems use heliostats to track the sun by two axis mechanisms following the azimuth and elevation angles to reflect the sunlight from many heliostats oriented around a tower and concentrate it towards a central receiver, situated atop the tower (see Figure 5.1). This technology has the advantage of transferring solar energy very efficiently by optical means, and of delivering highly concentrated sunlight to one central receiver unit, serving as energy input to the power conversion system.

In spite of the elegant design concept and the future prospects for high concentration and high efficiencies, the central receiver technology still needs some research and development efforts and demonstration of scaled-up plant operation to come up to commercial use. Most currently available central receiver technologies will require only one intermediate step before full commercialization. The main attraction is in the prospect of high process temperatures, generated by highly concentrated solar radiation, supplying energy to the topping cycle of any power conversion system, and feeding effective energy storage systems that are able to cover the demand of modern power conversion systems.

The solar thermal output of central receiver systems can be converted to electric energy in highly efficient Rankine cycle/steam turbine generators, in Brayton cycle/ gas turbine generators, or in combined-cycle generators (gas turbine with bottoming steam turbine). Grid-connected tower power plants are applicable up to about 200 MW$_e$ solar-only unit capacity. Conceptual designs of power plant units above 100 MW$_e$ have been analysed for ISCCS plants.

5.2.2.1 Energy service sector and role of technology

The technical feasibility of the central receiver (tower) technology has been proven worldwide between 1981 and 1986 by the operation of six research or proof-of-concept solar power plant units ranging from 1 to 5 MW$_e$ capacities, and by one pilot demonstration plant (SOLAR ONE with water/steam receiver) connected to the Southern California grid, totalling to a net electric capacity of 21.5 MW$_e$ with an installed heliostat mirror area of about 160,000 m². However, central receiver technology has not been used commercially up to now.

The 10 MW$_e$ SOLAR ONE pilot demonstration plant was operated in California from 1982 to 1988 with steam as the heat transfer medium. Rebuilt into the 10 MW$_e$ SOLAR TWO plant, it was successfully operated with a molten salt-in-tube receiver system and

Figure 5.4 SOLAR TWO 10 MW$_e$ central receiver power plant at Barstow, California, USA. (Source: Sandia National Laboratory, Albuquerque, USA)

a two-tank molten salt storage system from 1997 to 1999 (Figure 5.4), accumulating several thousand hours of operating experience and delivering power to the grid on a regular basis. SOLAR TWO has successfully demonstrated grid-connected operation and dispatching of the generated energy as designed. This concept is the basis for the US efforts for tower plant commercialization; it has the potential of more than 15% annual solar-to-electric plant efficiency and of annual plant availability of over 90%.

Different receiver heat transfer media have been successfully researched (water/steam, liquid sodium, molten salt, ambient air). Two of the proof-of-concept power plants have been converted to central receiver test bed facilities at the largest European solar test centre, the Plataforma Solar de Almería (PSA) in Spain (Figure 5.5), and two further large test facilities are available for R&D activities in the USA and in Israel. Today, promising advanced systems refer to the European volumetric air receiver technology and the US molten salt-in-tube receiver technology; both concepts are assessed to be the most mature and promising central receiver technologies for mid- to long-term grid-connected power plant applications. Heliostat technology, which is already available at near-to-commercial conditions in Europe (Germany, Spain) and in the USA, comprises glass–metal and stretched membrane heliostat types of 90–150 m² reflective surface area.

More recent European activities have demonstrated that high-flux characteristics are maximized in high-intensity/high-efficiency volumetric air receivers in either open or closed cycles, in which the concentrated solar energy irradiates fine wire mesh or ceramic foam or honeycomb structures, transferring the energy by convection directly

Figure 5.5 CESA I tower test facility at the Plataforma Solar de Almería. (Source: CIEMAT-PSA, Spain)

to air in the attractive temperature range of 700–1200°C. Tests conducted at the PSA in a joint German/Spanish project between 1993 and 1995 within the PHOEBUS Technology Programme Solar Air (TSA) with the German 2.5 MW_{th} pilot experimental plant showed the feasibility and prospects of the volumetric air receiver system concept with a ceramic energy storage system.

Plants may be designed as solar only for peak load (with about 2000 equivalent full-load hours per year) or in the future for up to 100% solar share on an annual average. Hybrid plant concepts, which are favourable for the phase of market awareness and expansion, are ISCCS plants for mid-load or base-load operation.

Future solar-only solar tower plants offer the good long-term prospect of high conversion efficiencies and the use of very efficient energy storage systems by utilizing high temperatures to enlarge the solar capacity or solar share. They also have the potential to be applied to other high-temperature heat processes (e.g. process heat, solar-chemical processes).

5.2.2.2 Technology shortages
Central receiver systems have the following technology shortages:

- No successfully scaled-up central receiver plants are yet available for commercial demonstration, although more than six experimental and pilot demonstration central receiver plants have been successfully operated worldwide since 1981.
- Currently promising technologies (the molten salt-in-tube receiver technology in USA and the volumetric air receiver technology in Europe, both with energy storage

system) are proven only by one pilot demonstration unit (10 MW$_e$ SOLAR TWO) with two years of operating experience and by one pilot experimental unit (2.5 MW$_{th}$ PHOEBUS-TSA) with some years of operating experience.
- The good potential projected for the improvement of solar system performance and for cost reductions is not yet verified.
- Projections of the installed plant capital costs, operation and maintenance costs, electricity costs, solar subsystem performance, operational characteristics and annual plant availability are not yet verified.
- Industrial volume production of heliostat components has still not been demonstrated.

5.2.2.3 Current projects
European activities related to the technical and economic feasibility of central receiver demonstration power plants have already started. Two 10 MW$_e$ projects are currently under development to take advantage of the Spanish Royal Decree of December 1998, which set up special legal classifications of premium payments for renewable or solar electricity, and of subsidies for solar investment from the EC Fifth Framework ENERGY Programme in its 1999 call for proposals. The projects in Spain are as follows:

- Two EU-funded central receiver projects, SOLGAS and Colón Solar, for a European demonstration of hybrid solar tower power plants (ISCCS) for integration of 20 MW of solar saturated steam into a conventional combined-cycle power plant; terminated after the detailed engineering phase in 1998. Their optimized integration scheme is awaiting future clarification of hybrid solar plants in Spain's legal framework to promote commercial demonstration plants.
- Planta Solar (PS10) 10 MW$_e$ solar-only power plant project at Sanlúcar near Sevilla, promoted by Spanish and German companies (EU co-funded); application of German PHOEBUS volumetric air receiver/energy storage technology; use of Spanish 90 m^2 glass–metal heliostats (Figure 5.6).

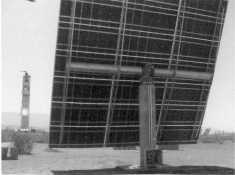

Figure 5.6 Spanish 90 m² glass–metal heliostat, Sanlucar-90. (Source: CIEMAT-PSA, Spain)

- 12.5 MW$_e$ solar-only power plant project with 16 h heat storage and 2.6 solar multiple (24 h operation strategy) at Córdoba promoted by Spanish and US companies; application of US molten-salt technologies for receiver and energy storage; use of new Spanish low-cost heliostats with reduced dimensions (no EU funding at present).

In addition, the following R&D projects concerning central receiver components or systems are being promoted or are in progress.

SPAIN
- SOLAIR project with 3 MW$_{th}$ high-temperature volumetric air receiver system at the PSA proposed by European group under Spanish industrial leadership; application for next central receiver projects after PS 10; with co-funds of the EC Fifth Framework Programme.
- REFOS project with a 1 MW$_{th}$ cluster of three closed pressurized volumetric air receivers at the PSA proposed by DLR together with CIEMAT (Figure 5.7); application for solar preheating of gas turbine combustion air in combined-cycle plants.
- SIREC project led by CIEMAT and IAER together with the Spanish company INABENSA; new low-cost heliostats with advanced technologies and a multi-layer knit-wire mesh volumetric receiver.

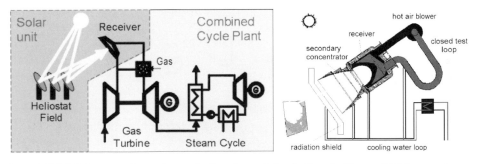

Figure 5.7 REFOS receiver for testing at the Plataforma Solar de Almería. (Source: DLR-Stuttgart)

ISRAEL
- DIAPR project by Israeli industry with German and US co-operation at central receiver facility of Weizmann Institute of Science (WIS) with tower reflector; for solar thermal or thermal-chemical applications.
- SOLASYS project by Israeli company with German co-operation at WIS central receiver test facility; application for synthetic gas production by solar reforming of natural gas; use of 400 kW$_{th}$ high-temperature volumetric air receiver (DLR).

5.2.2.4 Competitors, cost situation and vision of cost range
Competitors are the current conventional grid-connected fossil-fuel-fired power plant technologies, particularly the modern natural-gas-fired combined-cycle plants in mid- or base-load operation mode.

Specific installed capital costs of built central receiver pilot plants are high. There are no electricity costs of commercial scaled-up plants available today. Economic analyses show more cost uncertainties compared with trough plants as the technology is less mature, but similar installed costs are being estimated for the first demonstration plants in Spain (below €2700/kW), and there is a high potential for cost reduction. Central receiver plants will have considerable potential when applied at high temperature and with energy storage systems.

As analysed by the World Bank, the capital costs of short-term central receiver plants are expected to range from €4300/kW$_e$ (next 130 MW$_e$ ISCCS plant with 30 MW$_e$ solar-generated capacity with storage) to €3300/kW$_e$ (next 100 MW$_e$ Rankine cycle plant with storage) for US/European construction price levels, with predicted total plant electricity costs of about €0.14–0.12/kWh.

In Europe, new short-term tower projects in Spain have demonstrated installed plant capital costs of the order of €2700/kW$_e$ for a central receiver power plant based on the Rankine cycle with a small energy storage system, and predicted total plant electricity costs of €0.20–0.14/kWh.

The actual range of installed heliostat field costs is €180–250/m² for small production rates in the USA, and €120–220/m² in Europe: a 15% discount on US/European price levels may be projected for developing countries as labour costs are lower. Heliostat field costs are expected to drop below €100/m² for high production rates in the long term.

Central receiver plant projects will in the future benefit from cost reduction effects similar to those mentioned for parabolic trough plants. If total plant costs for grid-connected central receiver plants evolve as expected, the total plant electricity costs will drop to €0.07/kWh in the medium term period (for a 50–100 MW$_e$ Rankine cycle plant or an ISCCS, both with storage) and to €0.05/kWh in the long term period (for a 200 MW$_e$ Rankine cycle plant with storage), as has been analysed by the World Bank. In the long term, central receiver plants are expected to produce electricity at costs approximately 25% lower costs than those for similar-sized trough plants. There is the promising long-term potential for Rankine cycle central receiver systems to compete with conventional peaking Rankine cycle plants (coal-fired, oil-fired) at good solar sites. As mentioned above for ISCCS trough plants, ISCCS central receiver plants are not expected to have lower electricity costs than conventional gas-fired combined-cycle plants.

5.2.3 Parabolic dish systems

A dish system uses a parabolic reflector to focus the sun's rays onto a dish-mounted receiver at its focal point (see Figure 5.1). A heat-transfer medium transfers the solar energy to the power conversion system, which may be mounted in one unit together with the receiver (e.g. receiver/Stirling engine generator unit) or on the ground. Thanks to its ideal optical parabolic configuration and its two-axis control for tracking the sun, the dish collector achieves the highest solar flux concentration and therefore the highest performance of all concentrator types in terms of peak solar concentration and of system efficiency. However, these collector systems are restricted to unit capacities of some 10 kW$_e$ for geometrical and physical reasons.

Dish technology is applicable to off-grid power generation, i.e. at remote places or at island situations. Dish systems may optionally be arranged in large dish arrays in order to accumulate the power output from the kW$_e$ capacity up to the MW$_e$ range. Further R&D and demonstrations will be needed before such systems can be introduced onto the market.

Most dish systems use a Stirling engine generator for power conversion, but alternatives are a water/steam-powered turbine, a piston engine generator, or a gas turbine generator. Peak loads for solar-only operation or for solar/fossil 'hybrid' operation with solar shares can average 50–100% annually.

Short- and medium-term turnkey dish/Stirling systems are projected with the option of hybrid operation, i.e. with supplementary combustion of natural gas integrated into the receiver component. Such systems are currently under development, and should be available for first demonstration projects soon.

5.2.3.1 Energy service sector and role of technology
Since the end of the 1970s, various experimental, prototype and demonstration projects worldwide have demonstrated the technical feasibility of small power systems for off-grid solar thermal electricity generation using parabolic dish units in the 5–25 kW$_e$ range. High conversion efficiencies are possible with dish/Stirling systems because of the high process temperatures that can be used in the Stirling engine. The record energy yield so far has been achieved by a 25 kW$_e$ US dish/Stirling system with a solar-to-electric system efficiency of 30%. Current development is mainly in Europe (Germany) and the USA, with promising unit capacities of 10 kW$_e$ and 25 kW$_e$ repectively. These systems are under proof-of-reliability operation in the USA and in southern Spain. Stretched-membrane dish concentrators, currently under proof-of-reliability testing at the PSA using German advanced technologies, holds great promise for further cost reductions to make them competitive with diesel stations in remote areas or on islands. The tools required for small-series production of 100 units per year are also being developed.

5.2.3.2 Technology shortages
Parabolic dish systems have the following technology shortages:

- The electricity output of a single dish/Stirling unit is limited to small ratings of e.g. 25 kW$_e$ for geometric and physical reasons. (The exception is a large Australian dish designed for use with a 50 kW$_e$ steam engine or turbine generator.)
- Large-scale deployment has not occurred yet.
- No commercial demonstration has been performed to date.
- Not yet demonstrated or verified are projections of capital costs, operation and maintenance costs, electricity costs, system performance or annual plant availability over the long run.

- The predicted potential for improvements of solar system performance and of cost reductions is still to be verified.
- Hybrid systems have inherently low combustion efficiency, and have still to be proven.
- No adequate energy storage system is applicable or available.
- Industrial-scale volume production of the dish components and Stirling engines will be needed for entry into appropriate market segments.

5.2.3.3 Current projects

The following R&D and demonstration projects are being promoted or developed:

- **Europe.** First industrial demonstration programme for continuous operation of six German 9–10 kW$_e$ dish/Stirling pre-commercial units of Schlaich, Bergermann & Partner (SBP) at the PSA (three DISTAL I-systems since 1992 and three DISTAL II-systems since 1997, Figure 5.8); over 30,000 operating hours accumulated by the DISTAL I-systems so far; promising advanced heat pipe receiver types and Stirling engines currently under development and testing for proof of system reliability; new 9–10 kW$_e$ dish/Stirling units under way for testing at the PSA within the EuroDish R&D programme with EU co-funds since 1998 (goal: cost reductions by advanced structures for commercialized European dish/Stirling systems)
- **Spain.** Feasibility study and small demonstration project promoted by a Spanish group in collaboration with Stirling Energy Systems (SES) consortium using a 25 kW$_e$ McDonnell Douglas (MDAC) unit for erection in the south-east of Spain.
- **USA.** First industrial series of five 25 kW$_e$ US second-generation prototype systems for extended testing in the south-west USA; projects will possibly be shortened or stopped because of reduced public R&D funding in the short term.

Figure 5.8 Six 9 kW$_e$ dish/Stirling units (projects DISTAL I and DISTAL II) under continuous proof-of-reliability operation at the Plataforma Solar de Almería. (Source: CIEMAT-PSA, Spain)

- **Australia.** First 400 m² pilot experimental 'big dish' project with up to 150 kW$_{th}$ capacity under scientific testing at the Australian National University (ANU) since 1994; designed for power generation using a 50 kW$_e$ steam engine generator or for cogeneration applications by solar steam production; alternative to the small-unit philosophy described above.

The Australian government is currently funding a 2.6 MW$_{th}$ solar power plant consisting of 18,400 m² 'big dish' units that will inject solar-generated steam directly into the steam turbine of an existing coal-fired power station. Another 400 m² dish collector unit was recently sold to the Israeli solar test centre in the Negev desert for solar R&D testbed purposes.

5.2.3.4 Competitors, cost situation and vision of cost range

Competitors for dish/Stirling systems are conventional small-scale off-grid generation systems with unit ratings of the kW$_e$ range up to about 10 MW$_e$ in peak- or mid-load operation at remote places: i.e. gas-, oil- or heavy fuel oil-powered diesel engine generators and PV systems, particularly in developing sunbelt countries and on islands with relatively high fuel costs.

Installed dish collector costs have reduced dramatically:

- 1982: €1250/m² (40 m² Shenandoah, USA)
- 1985: €300/m² (91 m² MDAC, USA)
- 1986: €200/m² (44 m² LaJet, USA)
- 1992: €150/m² (44 m² German SBP stretched-membrane dish).

Current installed plant capital costs for a first stand-alone 9–10 kW$_e$ dish/Stirling unit are €10,000–14,000/kW$_e$ and for actual short-term units €7100/kW$_e$ (at a production rate of 100 units/yr). A realistic short-term goal for electricity costs is less than €0.15/kWh. In the medium to long term, installed system costs are expected to decrease dramatically as series production of dish units increases. The goal of the European EuroDish project is for costs to drop as follows:

- €7100/kW$_e$ (100 units/yr)
- €3700/kW$_e$ (1000 units/yr)
- €2400/kW$_e$ (3000 units/yr)
- €1600/kW$_e$ (10,000 units/yr).

Further cost reduction is unlikely because of the inherently very high modular technology. In the medium to long term, installed dish collector costs are predicted in the range €125–105/m² for high production rates, so that advanced dish/Stirling systems have the potential to compete with similar-sized diesel generator units in sunny remote places or islands.

5.3 GOALS FOR RESEARCH, DEVELOPMENT AND DEMONSTRATION (RD&D)

The main goals for solar thermal power plants are to demonstrate commercial-scale projects, and to reduce the cost of components and subsystems.

5.3.1 Commercial-scale projects

The top priority is to erect, in the Mediterranean area, grid-connected *large solar thermal power plants (hybrid or solar only), with significant solar shares and commercial operation*, and minimum equivalent solar capacities of 10 MW$_e$. This will ensure a significant contribution to environmentally benign electricity production in Europe. In this respect, Spanish and Greek initiatives will play a decisive role, enjoying the support of the European Union.

Integrated hybrid solar/fossil solutions (ISCCS plants) have to be followed up. They link proven solar thermal and conventional technologies through optimized integration of solar thermal energy into fossil-fired power plants, particularly combined-cycle power stations. Such applications are not just door openers for solar applications; they represent a realistic strategy. Other schemes for optimized integration of central receivers enjoy EC support, such as the demonstration of small, modular integrated multi-tower systems operating on an advanced Ericsson cycle, and the integration of central receivers with biomass.

5.3.2 Cost reduction

An essential objective for R&D is to *reduce the cost of improved and optimized components and subsystems*, as follows:

- Qualify and demonstrate solar direct steam generation in a multiple-row parabolic trough solar field.
- Develop cost-reducing mass production techniques for absorber tubes with highly efficient solar coatings, with and without vacuum.
- Develop mass production techniques to reduce the fabrication and assembly costs of the concentrating reflector structures for all solar thermal technologies.
- Develop low-cost, fully autonomous heliostats with improved communications, advanced drives and self-structuring facets.
- Develop new absorber coatings with economical mass application procedures.
- Develop reliable, high-flux, compact receiver systems.
- Integrate fossil backup solutions into the receivers of power towers and dish/Stirling systems.
- Develop new, environmentally benign procedures for reflector cleaning.
- Further reduce the cost and improve the performance of thermal storage systems for all solar thermal technologies.
- Develop control concepts for unattended and remote operation.
- Integrate solar thermal technologies into organic Rankine cycles.

- Identify long-term high potential photo-electric direct conversion systems.
- Develop graphical, object-oriented simulation software tools for optimizing physical, economic, financial and environmental performance.
- Develop advanced measurement systems for continuous monitoring of the efficiency of the solar energy conversion path of large-scale central receiver and parabolic trough plants, in order to optimize their operation.

5.4 ROADS TO THE MARKET

Together with hydro power stations and large wind converter arrays, solar thermal power plants have the potential to generate competitively priced bulk electric power, and thereby contribute to international efforts to reduce climatic gas emissions from fossil-fired electricity generation. They can also utilize the proven high-voltage direct current transmission system in order to link good solar sites in the Mediterranean, via grid connection, with consumers of high electricity demand in less sunny locations.

All three solar thermal power plant technologies are nearly ready for the market, although they are at different stages of technological and commercial availability: parabolic trough technology ready now, with further improvement from direct steam generation in the short term period, and the central receiver and dish/Stirling technologies in the short to medium term.

Figures 5.9 and 5.10 show respectively the status of international developments and the evolution of the levelized energy costs of parabolic trough and central receiver power plant projects.

The GEF's operational programme provides financial support for specific solar thermal power applications in sunbelt countries, as mentioned above.

The roads to the market should be pursued step by step: first the rapid market introduction of proven SEGS-type parabolic trough plants, then implementation of a limited series of commercial central receiver power plants, and finally the use of dish/Stirling applications in appropriate market segments.

The first step for all solar thermal power technologies is to identify the most probable market segments through market analysis, awareness and acceptance. The technologies then have to be demonstrated by stages, with the final goal of commercial introduction within the next decade. As for solar/fossil hybrid solutions, solar systems should be demonstrated by coupling them to conventional power plants in order to optimally combine the technical concepts and their economies, particularly during the market introductory phase. However, these recommendations should be taken with a certain measure of flexibility so that they remain adaptable to circumstances and do not become intolerant of local requirements. Recent experience in Spain, the USA and Australia demonstrates that open-minded strategies are necessary to implement solar thermal power plants successfully. In a world in which the recognition of differences is increasingly important, the only viable strategies for introducing a new technology are those that integrate the user or receiver of the technology in the defining the strategy itself, and which respect social circumstances and the expectations of those for whom the solution is proposed.

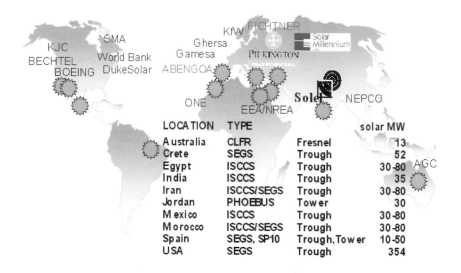

LOCATION	TYPE		solar MW
Australia	CLFR	Fresnel	13
Crete	SEGS	Trough	52
Egypt	ISCCS	Trough	30-80
India	ISCCS	Trough	35
Iran	ISCCS/SEGS	Trough	30-80
Jordan	PHOEBUS	Tower	30
Mexico	ISCCS	Trough	30-80
Morocco	ISCCS/SEGS	Trough	30-80
Spain	SEGS, SP10	Trough,Tower	10-50
USA	SEGS	Trough	354

Figure 5.9 Status of international developments of solar thermal power plant projects. (Source: DLR-Almeria

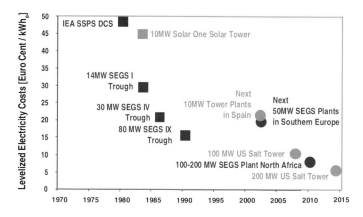

Figure 5.10 Evolution of levelized electricity costs for solar thermal power plants referred to solar-only production for sites with different insolation. (Source: DLR-Almeria)

5.4.1 Parabolic trough systems

Steps along the road to the market will be as follows:

- building on the success of current parabolic trough collector systems and taking a low-risk approach to advance the technology during the next decade, in order to

reach the necessary synergy of technology development and market awareness, expansion and acceptance
- hybrid ISCCS plants using current trough technology; hybrid plant design optimization and optimized linkage to conventional power plants or to co-generation plants.
- improvements of the key elements of trough collectors, e.g. optimized steel structures, absorber tubes with increased lifetime and better maintainability, special operation and maintenance tools and equipment, improved procedures
- standardized designs for hybrid ISCCS cycles with equivalent solar unit capacities of 30–50 MW$_e$ and for solar Rankine cycles with capacities from >30 MW$_e$ to 200 MW$_e$
- advanced trough collector/reflector designs, e.g. mirror facets with front surface or film reflectors of high reflectivity, lighter and strengthened mirror panels
- demonstrating direct steam generation trough technology ready for commercialization.
- developing and applying efficient energy storage systems in future solar thermal power plants in order to enlarge the solar capacity of mid-load or near-base-load plants with average annual solar shares up to 100%.

5.4.2 Central receiver systems

Steps along the road to the market will be as follows:

- projecting the success of current central receivers and taking a low-risk approach to advance the state of the central receiver technology by several steps during the coming decades, in order to reach the necessary synergy of technology development and market awareness, expansion and acceptance
- hybrid ISCCS plants using current central receiver technology; hybrid plant design optimization and optimized linkage to conventional power plants or to co-generation plants and to biomass plants
- improving heliostat fields and receivers by better optical and thermal properties, lighter mirrors, optimized heliostat structures, and better heliostat/field controls
- proving solar system performance and reliability during representative operating times
- process improvements through further advances in selected heat transfer media and receivers for commercial power generation (e.g. European volumetric air receiver technology, US advanced molten salt-in-tube receiver technology)
- improving system integration through reduced parasitic loads, optimized start-up procedures, better control strategies, automatic heliostat field control
- developing and applying low-cost, very efficient energy storage systems in future central receiver power plants in order to enlarge the solar capacity of mid-load or near-base-load plants with average annual solar shares up to 100%.

5.4.3 Parabolic dish systems

Steps along the road to the market will be as follows:

*   improvements of dish reflector and receiver, including better optical properties of the mirrors; lighter mirrors and structures, better system control characteristics; development of an automatic control system for remote operation and for long-distance control
*   system improvements using Stirling and Brayton (gas turbine) engines adapted to solar processes with advanced heat-pipe and volumetric air receivers
*   proof of reliable operation of advanced Stirling engine/receiver units over the long run
*   improvements in system integration by reduction of parasitic loads, optimization of start-up procedures, better control strategies and hybrid operation of Stirling or Brayton engines.

BIBLIOGRAPHY

Becker, M and Meinecke, W (eds), *Solar Thermal Technologies: Comparative Study of Central Receiver, Parabolic Trough, Parabolic Dish and Solar Chimney Power Plants* (in German). Springer, Berlin, 1992.

Becker, M and Gupta, B (eds); Meinecke, W and Bohn, M (authors), *Solar Energy Concentrating Systems, Applications and Technologies*. C F Müller, Heidelberg, 1995.

Becker, M and Böhmer, M (eds), *Proceedings of the 8th SolarPACES International Symposium on Solar Thermal Concentrating Technologies, Köln, 1996*. C F Müller, Heidelberg, 1999.

Blanco, M and Ruiz, V, The SOLGAS cogeneration project: a paradigm shift for the solar thermal industry. *Proceedings of the 7th SolarPACES International Symposium on Solar Thermal Concentrating Technologies*, Moscow, 26–30 September 1995.

Chavez, J M, Kolb, G J and Meinecke, W (authors), *Second Generation Central Receiver Technologies: A Status Report*. C F Müller, Karlsruhe, 1993.

Cohen, G E, Kearney, D W and Kolb, G J, *Final Report on the Operation and Maintenance Improvement Program for Concentrating Solar Power Plants*. SANDIA Report SAND99-1290, Albuquerque, June 1999.

Enermodal Engineering Ltd, *Cost Reduction Study for Solar Thermal Power Plants: Final Report*. Prepared by Enermodal Engineering Ltd in association with Marbek Resource Consultants Ltd, by contract of World Bank/GEF, Washington DC, 5 May 1999.

Flamant, G, Ferriere, A and Pharabod, F (eds), Proceedings of the 9th SolarPACES International Symposium on Solar Thermal Concentrating Technologies, Font-Romeu, June 1998. *Journal de Physique IV*, Vol. 9, 1999.

Geyer, M and Klaiß, H, 194 MW$_e$ solar electricity from trough collector plants (in German). *BWK Journal*, Vol. 6, 1998, pp. 288–295.

Geyer, M, Holländer, A, Aringhoff, R and Nava, P, 7000 GWh solar electricity from parabolic trough power plants, half of world-wide produced solar electricity (in German). *Sonnenergie*, Vol. 3, June 1998.

Klaiss, H and Staiss, F (eds), 1992, *Solar Thermal Power Plants for the Mediterranean Region* (in German), Vols I and II. Springer, Berlin, 1992.

Pilkington Solar International (eds), *Solar Thermal Power – Now, A Proposal for Rapid Market Introduction of Solar Thermal Technology*. Co-operative Position Paper of Pilkington Solar, Kramer

Junction Co., SOLEL, DLR, Plataforma Solar de Almeria, CIEMAT, Bechtel, Fichtner and Schott Rohrglas, 1996.

Stine, W B, and Diver, R B, *A Compendium of Solar Dish/Stirling Technology*. SANDIA Report SAND93–7026 UC-236, Albuquerque, January 1994.

Trieb, F, Competitive solar thermal power stations until 2010: The challenge of market introduction (SYNTHESIS Programme). Proceedings of the 9th SolarPACES International Symposium on Solar Thermal Concentrating Technologies, Font-Romeu, June 1998. *Journal de Physique IV*, Vol. 9, 1999.

Tyner, C, Kolb, G J, Meinecke, W and Trieb, F, *Concentrating Solar Power in 1999: An IEA-SolarPACES Summary and Future Prospects. SolarPACES Task I: Electric Power Systems*. SolarPACES, Albuquerque, January 1999.

LIST OF ABBREVIATIONS

ANDASOL	32MW parabolic trough project development in the province of Almería
CESA 1	1 MW$_e$ Central Electrosolar Uno at the PSA, Tabernas, Spain
Colon Solar	Co-generation project using central receiver technology at Colon, Spain
DIAPR	Directly irradiated annular pressurized receiver
DISS	Direct solar steam project
DISTAL	Dish/Stirling Almeria, R&D project at the Plataforma Solar de Almería, Spain
EuroDish	European dish/Stirling R&D programme
EuroTrough	European parabolic trough R&D programme
ISCCS	Integrated solar combined-cycle system
KfW	Kreditanstalt für Wiederaufbau, Germany
LEC	Levelized electricity costs
LS-3	LUZ System, parabolic trough system model No. 3 of LUZ
PHOEBUS-TSA	PHOEBUS technology programme solar air receiver
PS 10	10 MW$_e$ central receiver power plant project Planta Solar 10 near Sevilla, Spain
REFOS	Modular pressurized volumetric air receiver, R&D project for solar preheating of combustion air of fossil-fired gas turbines and combined-cycle power plants
SEGS	Solar electric generating systems
SIREC	Sistemas de receptor central (systems for central receiver) project, involving CIEMAT and IAER for heliostat and receiver development, Spain
SOLAIR	Advanced solar volumetric air receiver for commercial solar tower power plants, European R&D project
SOLASYS	Novel solar-assisted fuel-driven power system
SOLAR ONE	10 MW$_e$ central receiver power plant No. 1, Barstow, CA, USA
SOLAR TWO	10 MW$_e$ central receiver power plant No. 2, Barstow, CA, USA
SOLGAS	Solar/gas-fired hybrid co-generation plant project using central receiver technology in Andalusia, Spain
THESEUS	50 MW$_e$ parabolic trough power plant project, Frangocastello on Crete, Greece

6. WIND ENERGY

J Beurskens (ECN), P H Jensen (Risø)

6.1 INTRODUCTION, POTENTIAL AND BASIC STRATEGY

6.1.1 Introduction

Wind energy is the kinetic energy of moving air. Air moves from high-pressure to low-pressure areas. The non-uniform heating of the earth's surface by the sun is the driving force behind the global wind system. The power of the wind is proportional to the cube of the wind speed. See Figure 6.1. This non-linear relationship, coupled with the fact that the wind varies constantly in magnitude and direction, is one of the fundamental reasons for the complexity of the technology and its application in electricity supply systems.

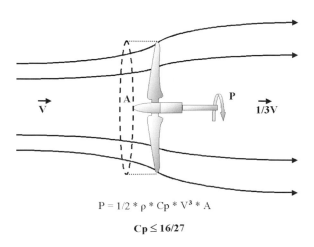

$$P = 1/2 * \rho * Cp * V^3 * A$$

$$Cp \leq 16/27$$

Figure 6.1 The power in the wind is proportional to the cube of the wind speed. The maximum fraction a wind turbine can extract from the wind is 16/27 (0.593).

The energy density of the wind is relatively low compared with densities in machines such as gas and steam turbines. Average power densities in high potential areas are typically 400–800 W/m² swept rotor area. These low densities are the reason why many large machines have to be built in order to generate significant amounts of energy.

6.1.2 Potential: the European wind energy resources

6.1.2.1 Resource assessment

When assessing the European wind resource, one has to look at the wind resources both on land and offshore. Since the early 1980s, land resources have been investigated and mapped. A large volume of data is available in the public domain[1] but an increasing amount of data is considered as strategic and confidential. Although offshore applications have been of interest since the late 1970s,[2] the first more or less comprehensive glimpse of the European offshore resource was provided in 1995 by an EC study called *Study of Offshore Wind Energy in the EC*.[3]

The local potential depends very much on the average wind speed, as the power in the wind is proportional to the cube of the (momentary) wind speed. For the design of an optimized wind turbine system, the statistical distribution of the wind speeds also has to be available for both energy output predictions and establishing design criteria. Thus, a thorough evaluation of the circumstances at the site is needed in addition to the overall averaged data. For offshore applications essential data – such as simultaneous wind speed and wave height data – are still lacking.

The technology and software (e.g. *The European Wind Atlas*[1] and the computer code WasP[4]) for evaluating the available wind resource are being used intensively by project developers. See Figure 6.2. The European Commission has played a key role by funding the research leading to the development of these tools. Despite the fact that in several other regions of the world evaluations of the wind resources have been carried out, a number of areas remain to be mapped. Although a lot of effort has been invested in resource assessment and wind turbine output prediction methods, the uncertainty of the results is still too high for economic assessments by energy economists, who are used to evaluating fossil fuelled plants with a high degree of dispatchability.

There are two reasons for the uncertainty in estimating the annual energy output:

- the annual variation of the average wind speed compared with the long-term average (in coastal areas the annual wind speed may vary by up to plus or 7% compared to the 30-year average)
- the inaccuracy of measurements and the use of those measurements for energy output calculations (terrain effects, inaccuracies of the wind turbine power characteristics) – currently, on average the uncertainty in the latter part of the calculations is of the order of 10%.

Figure 6.2 Example of a wind resource map from The European Wind Atlas.

	Sheltered terrain[2]		Open plain[3]		At a sea coast[4]		Open sea[5]		Hills and ridges[6]	
	m s[-1]	Wm[-2]	m s[-1]	Wm[-2]	m s[-1]	Wm[-2]	m s[-1]	Wm[-2]	m s[-1]	Wm[-2]
	> 6.0	> 250	> 7.5	> 500	> 8.5	> 700	> 9.0	> 800	> 11.5	> 1800
	5.0-6.0	150-250	6.5-7.5	300-500	7.0-8.5	400-700	8.0-9.0	600-800	10.0-11.5	1200-1800
	4.5-5.0	100-150	5.5-6.5	200-300	6.0-7.0	250-400	7.0-8.0	400-600	8.5-10.0	700-1200
	3.5-4.5	50-100	4.5-5.5	100-200	5.0-6.0	150-250	5.5-7.0	200-400	7.0-8.5	400-700
	< 3.5	< 50	< 4.5	< 100	< 5.0	< 150	< 5.5	< 200	< 7.0	< 400

Wind resources[1] at 50 metres above ground level for five different topographic conditions

6.1.2.2 The wind resources on land

Areas potentially suitable for wind energy application are dispersed throughout the European Union. Major areas that have a large wind energy resource include Great Britain, Ireland and the north-western continental parts of the EU: Denmark, northern Germany, south-western Sweden, the Netherlands, Belgium, and north-western France. Other areas include large parts of Spain, a majority of the Greek islands, and some parts of Italy. In addition, there are many areas where wind systems associated with mountain barriers give rise to high energy potentials. Some of these wind systems extend over large areas: the Mistral between the Alps and the Massif Central in the South of France, and the Tramontana north of the Pyrenees in France and south of the

Pyrenees in the Ebro valley. In other cases such wind systems are of smaller geographical extent, but may nevertheless present a large wind resource locally. Of particular interest are mountain valleys, funnels and ridges where natural concentration effects of wind occur (e.g. mountainous parts of Germany, Austria, Portugal, Italy, Sweden and Finland).

6.1.2.3 The offshore resources

The scope of the offshore study[3] mentioned above encompassed the assessment of the total offshore wind resource of Europe (12 countries), gathering experience from the offshore industry concerning construction, erection, transportation and maintenance relevant to offshore wind farm operation and design aspects and adaptation of computer codes accounting for the differences between land-based and offshore machines (combined wind and wave loading, etc.).

The study showed Europe's technical offshore potential to be greater than its total electricity consumption. See Figure 6.3.

These estimates were based on the selection of sites having a water depth between 0 and 40 m, a distance to the shore not exceeding 30 km and a generating density of the wind turbines of 6 MW/km². The authors state that the review most probably overestimates the exploitable resource because political and other constraints relating to alternative use of the sites were not taken into account. On the other hand, there are considerable resources outside the 30 km limits, for example in the North Sea.

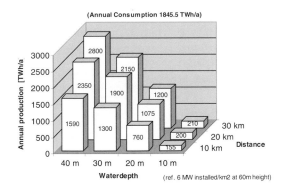

Figure 6.3 The European offshore potential (12 countries). Bars indicate cumulative figures

6.1.2.4 The total resource of Europe

The overall conclusion that can be drawn from the resource information available is that the total exploitable wind resource in the countries of the European Union is more than twice the total present electricity demand.

6.1.3 Basic strategy

Various (international) governmental bodies and professional organizations have formulated goals to exploit the wind resource in Europe. Besides meteorological data, these goals are based on infrastructure, physical planning, and environmental and economic considerations. Apart from these issues economic development, employment and decreasing the tension between the industrialized 'north' and the developing 'south' provide important arguments for a policy to stimulate the introduction of renewable energy sources in general and wind energy systems in particular.

The European Wind Energy Association (EWEA) along with Greenpeace concluded that even if the demand for electricity increases quickly (according to the IEA the world electricity demand will double by 2020), wind energy will be able to provide 10% of the world demand by the year 2020.[5] This implies an average annual growth of installed wind power of 20% between 1998 and 2003 (at present the growth is about 30%), 30% between 2004 and 2010 (resulting in 181,000 MW) and 20% between 2010 and 2020 (resulting in 1,200,000 MW), generating 3000 TWh, equivalent to almost 11% of world electricity consumption.

In the past EWEA was twice forced to adjust its targets to higher figures because of the spectacular growth of sales of wind turbines in the European Union. Its present targets are as follows: 8000 MW by 2000 (at the end of 2000 the actual figure was approximately 13,600 MW!), 60,000 by 2010 – of which 5000 MW is to be installed offshore – and 150,000 MW by 2020.

In order to meet these ambitious goals a number of conditions have to be fulfilled. Essentially this means a continuation of the success of the three countries on the European continent that now host 82% of the total installed power in Europe: Denmark, Germany and Spain (worldwide five countries constitute 82% of the total installed power: Denmark, Germany, India, Spain and the USA). Analysing the success factors of those countries is crucial in order to 'repeat' these developments and guarantee an accelerated market growth. Of course strategies have to be adapted to the specific circumstances in the countries concerned.

In general, the following can be considered as success factors:

- **Cost reduction** of wind energy.
- **Improving the value of the wind electricity.** In the first instance value is determined by the avoided fuel cost of fossil-fuelled plants. Equally relevant are cost components that are seldom made explicit in the electricity market, such as the environmental advantages, capacity credit (to be improved by utilizing methods to forecast the output of wind farms a fewer hours in advance), and demand for green electricity by customers.
- Finding **new sites**. In densely populated coastal countries – offshore; in mountainous land locked countries – funnels, hills, mountains.
- **System development**. On the one hand national grids have to be able to absorb large amounts of varying electricity output; on the other hand, wind energy plants need to meet specific requirements such as the amount of reactive power produced, harmonic distortion, predictability and controllability of the output.
- **Reduction of uncertainties** in predicting the technical and economic feasibility of projects. This means improved resource assessments, wind speed measurements, maintainability of machines, reliability, lifetime design methods.

- **Reducing environmental and negative social impacts** such as visual impact on the landscape, effects on birds and their habitats, acoustic noise emissions, etc.
- **Education and human resource development.** The EWEA and Greenpeace expect a workforce of 1.7 million if the 10% target for 2020 is to be met. People have to be educated and trained in both technical and non-technical capabilities. Training precedes actual employment!
- **Implementing national and international environmental and energy policies.** A large number of instruments have been implemented in the past mostly by national states. Some schemes proved to be very efficient and some were not. The proper integration of financial and other legal measures, in relation to the maturity of the technology, appeared to be a necessary condition. On the international stage a number of agreements are being developed to protect the environment and in which renewable energy plays a crucial role: Joint Implementation (JI) and Clean Development Mechanism (CDM).

The first two factors of course are closely linked. The bigger the difference between the cost of wind energy and its value, the greater the driving force behind the economic development of the wind energy technology.

Educational establishments such as universities and polytechnic schools, the R&D communities in the institutes and in industry have to provide the know-how and capabilities to further specify the success factors, and to develop and operationalize them.

The aspects discussed above are to be linked to strategic goals in order to be effective. The strategic goals encompass the future wind energy generating capacities in different markets, industrial and employment goals. Table 6.1 links the success aspects to these strategic goals. The number of bullets indicates the relative importance of the aspect concerned in order to meet the strategic goal. The importance depends, of course, on the maturity of the technology, the stage of development and know-how.

Table 6.1 The relations between strategic goals and aspects of implementation success factors for wind energy systems

Success factors	Strategic goals		
	10% renewable energy in Europe	30% new renewable energy capacity in developing countries	Maintaining industrial capacity and employment
Cost reduction of wind energy	••	•••	•••••
Increasing the value of wind energy	•••	•••	•••
Finding new sites	•••••	••	•
System development	•••••	•••••	•••
Reduction of uncertainties	••	•••••	•••••
Reduction of environmental effects and negative social impacts	•••	-	•
Education and human resource development	••	••	•••••
Development of policy and instruments	•••••	•••••	•••••

EWEA's wind energy targets form a subgroup of the European Union's target of covering 10% of Europe's energy demand by means of renewable energy. To emphasize the importance of renewables for the developing economies, a target has been arbitrarily added for those countries. This target again is a subgroup of EWEA and Greenpeace's recommendation of covering 10% of the world electricity demand by means of wind energy by the year 2020.

Later an R&D roadmap will be developed on the basis of the relation between success factors and types of wind energy systems that have to be commercially available in order to meet the goals, discussed above.

6.2 PRESENT SITUATION AND ACHIEVEMENTS

6.2.1 Status of the technology

Depending on the application, wind turbine systems exist in a range of capacities. Usually the technology is divided into three categories, according to Table 6.2.

Figures 6.4 to 6.9 represent typical examples of systems from the different categories.

Table 6.2 Wind turbine categories for different applications

	Type of application	Type of wind turbine
A	Land-based and offshore grid-connected wind farms varying in size from about 10 MW to several 100 MW.	Installed power > 1.5 MW
B	Decentralized and single operating machines connected to the grid.	Installed power < 1.5 MW
C	Decentralized units for grid-connected operation and for hybrid and stand-alone operation.	Installed power < 0.5 MW

Figure 6.4 Example of a category A application. A 2 MW wind turbine with a rotor diameter of 72 m, designed for offshore applications. Further characteristics: active stall control by means of limited blade pitch control, constant rotational speed. (NEG-Micon)

Figure 6.5 Category A. A wind farm consisting of 35 machines of 1.5 MW each and a rotor diameter of 66 m at Holtriem, Germany. Further characteristics: blade pitch control, variable rotational speed and direct drive generator. (ENERCON)

Figure 6.6 Example of a category B application. A Danish offshore wind farm at Tunø Knob, consisting of 10 machines of 500 kW each and a rotor diameter of 39 m. Further characteristics: blade pitch control and limited variable rotational speed. (ELSAM)

Figure 6.7 Category B. 'La MUELA III' wind farm near Zaragoza, Spain, comprising 26 machines of 660 kW each and a rotor diameter of 46 m. (MADE)

Figure 6.8 Category C. Single grid-connected 400 kW wind turbine with a rotor diameter of 31 m, owned and operated by a farmer at Sint Maartensbrug, the Netherlands. Characteristics: stall control (fixed rotor blades) and constant rotational speed. (Jos Beurskens)

Figure 6.9 Example of a category C application. An experimental wind–diesel system under test at ECN, Petten, the Netherlands. Characteristics: installed power of diesel set: 50 kW, wind turbine: 35 kW. Passive blade pitch control, variable rotational speed. (Jos Beurskens)

6.2.1.1 Category A

Category A is formed by large-size grid-connected wind turbines for land-based and offshore applications.

The size of commercially available grid-connected horizontal axis wind turbines has evolved from 22 kW in the early 1980s, via 0.6 MW in the early 1990s, to about 2.5 MW today (associated rotor diameters are approximately 10 m, 40 m and 90 m, respectively). The next generation of commercial wind turbines in the 2.5–5 MW range are being tested and are scheduled to reach the market in 2002 to 2005. The larger machines are designed mainly for offshore applications.

Many different design concepts are employed. There is a strong tendency for large machines to move from stall-regulated machines operating at near-fixed rotational speed to concepts with variable speed and blade pitch control. Other concepts with direct-driven generators or low-ratio gearboxes and integrated low-speed generators are occupying an increasing share of the market.

The European industry is leading the rest of the world in the development and manufacturing of grid-connected wind turbines. In 1999 the European industry produced slightly more than 90% of worldwide sales. However, as the ownership of the companies and the manufacturing facilities is more dispersed, it becomes increasingly difficult to determine the 'nationality' of the technology.

Modern wind turbines have a high quality, which results in technical availability of typically 98–99%. The high level of availability is a result of the reliability of the machines and – for most land-based machines – the easy accessibility for repairs and maintenance. However, under offshore conditions the situation is different. Because of poor accessibility of the site, a higher degree of reliability is one of the necessary conditions to be fulfilled in order to arrive at an acceptable level of availability. See Figure 6.10. Besides transportation and installation loads, combined wave and wind loads and reliability are very important design drivers for offshore turbines.

Category A (and also B) wind turbines require more fundamental research as dynamic, mechanical, aerodynamic and aeroelastic phenomena play a dominant role in creating stable constructions, whereas R&D on category C machines concentrates on energy system improvements.

More and more wind turbines are placed in clusters, so-called **wind farms**, which are operated as single energy-generating units. Unit size has evolved from a few MW in the early 1980s to hundreds of MW at present.

It is estimated that a penetration (supply fraction) of wind energy on a large grid can be as much as 15–20% without special precautions to secure grid stability. Turbines equipped with state-of-the-art power electronics may even add stability to the grid,

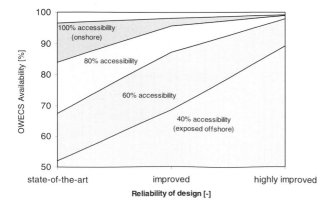

Figure 6.10 Availability of wind turbines in relation to reliability and accessibility of the terrain.[6]

because reactive power can be adjusted and the harmonic distortion of the electricity generated is extremely low. Also power fluctuations can be limited as the output of wind farms is partly controllable.

The fact that wind turbines are clustered into bigger units requires specific R&D on system integration. Focal points are: control systems of wind farms; improvement of power quality; system and grid stability; operational management systems (maintenance based on conditioning monitoring and early failure detection); system reliability; and output forecasting.

As land-based large wind farms are very visible and produce noise, these aspects among others play an important role in the planning process, especially in densely populated countries where the best locations are already occupied.

6.2.1.2 Category B

Category B is where most of the generating capacity of the first independent wind farms and later individual wind turbine owners can be found. The machines in this category are almost all installed on land. (Exceptions are the first demonstration offshore installations in Sweden and Denmark.) Generating capacities are matched to the local grid capacity (short circuit power). These machines are also being used in large wind–diesel units.

Predictions of future generating cost of wind turbines and long-term wind capacity targets (for instance, those from the EWEA and Greenpeace) are based upon the experience of wind turbines from this category. See Figure 6.11. By the end of 1999 almost 44,000 machines (including all categories) had been sold worldwide.[8]

Areas of improvement are power quality, controls and system design for autonomous hybrid units. As the machines in this category (and also the machines of 100–500 kW from category C) are the oldest ones in operation, wear and tear and fatigue-induced failures appear more frequently as the end of the projected lifetime comes nearer. This leads to increased maintenance cost in the overall statistics.

Questions of replacing those machines by larger state-of-the-art machines or refurbishing the old ones are being addressed by project owner/operators.

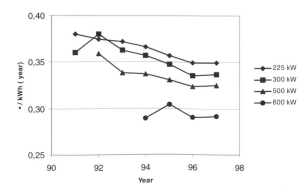

Figure 6.11 Cost development of medium-size wind turbine[7] in terms of ex-factory turbine list prices per unit of annually produced electricity (Roughness class 1; 1996 prices; €1 = DKK 7.5).

6.2.1.3 Category C

GRID-CONNECTED MACHINES

After the oil crisis in the mid 1970s, the commercial development of grid-connected electricity generating machines began in countries like Denmark, the Netherlands and the USA with machines of typically 22–35 kW installed power and rotor diameters of 10 m. Owners were almost exclusively private persons, farmers and small enterprises. For remote areas with weak grids in both the industrialized and developing world, machines between, say, 50 kW and 500 kW still find a sizeable market that is ignored by the established industry because of the booming market for larger machines. Some niche-oriented manufacturers are focusing on this market segment. A relatively new development is the use of small, grid-connected machines (smaller than 10 kW) in the built environment (UK) and farms (USA).

AUTONOMOUS SYSTEMS

In the early 1980s, at the beginning of the modern renewable energy development period, relatively much attention was paid to utilizing wind energy systems to provide electricity to remote, isolated communities without any energy infrastructure. As the supply of wind energy varies stochastically over time, energy supply and energy demand do not match. To provide security of supply, a storage system is needed that absorbs energy during periods in which supply exceeds demand and supplies energy at times when the situation is reversed. Storage of energy in general and of electricity in particular is very expensive. In order to overcome this cost problem solutions were sought in so-called autonomous – hybrid – systems where wind turbines were working in combination with other energy conversion systems such as photovoltaic arrays/hydro/diesel and/or storage systems. These wind turbines were also used for special stand-alone applications such as water pumping, battery charging, desalination, etc. (10–150 kW range).

Many experiments were carried out, and over 10 research establishments had more or less comprehensive development programmes. In retrospect, one can conclude that virtually all development efforts on autonomous systems had limited success.

The need for independent island renewable energy systems has not changed, however. Analysing and understanding the reasons why this happened could provide us with the conditions for a successful revival of the development and market implementation of autonomous systems. The generally accepted notion that the market did not develop is that the actual players (technology developers, manufacturers) were not familiar with the demand side of the market for energy supply to remote places. However, it also has to be considered that this market is not self-sufficient and lacks even the most basic technical and logistic infrastructure for after-sales service. While developing renewed development strategies, the aspects of the demand side of the market should be fully integrated, e.g. by involving the established manufacturers and distributors of diesel sets for remote electricity supply.

This category also constitutes **small stand-alone turbines** for battery charging, water pumping, heating, etc. (<10 kW). Small battery-charging wind turbines (25–150 W; i.e. rotor diameters from 0.5 m to 1.5 m) are by far the most successful in commercial terms. At present probably some one million small battery charging wind turbines are in operation worldwide.

Of all forms of applications of wind turbines, wind driven pumps still are the most numerously applied. One to two million units are used worldwide for irrigation, cattle watering and domestic supply. Over 50 manufacturers are known to be active. In general the technology has not developed significantly since the first machines appeared on the market in the mid 1800s. Few improvements from R&D activities supported by national programmes were implemented successfully. Interest in these machines seems to be increasing.

6.2.1.4 General

Although the market share of category B machines in 1999 was still the largest (about 70% of the total number of machines sold), the market share of category A machines is growing rapidly.

6.2.2 Markets[8]

6.2.2.1 General trends

Installed wind power is distributed very unevenly over the world: only five countries make up 82% of the total installed capacity (which was nearly 18,500 MW at the end of 2000). See Table 6.3. Also the annual growth of the capacity is unevenly distributed: in one year (2000) Germany installed about four times (1665 MW) the cumulative wind capacity of the Netherlands, although the exploitable wind resources in both countries are of the same order of magnitude.

Over the last 20 years the worldwide annual installation rate of wind turbines has increased from a few megawatts to almost 4500 MW in 2000. From 1995–2000 the growth of the cumulative capacity varied between 26% and 37% and the average growth was 31%. Europe hosts 74% of the world's capacity: 13,630 MW at the end of 2000. In the 1980s, for example, the generation cost of wind electricity was reduced approximately by a factor 10.

Table 6.3 *Installed world wind capacity 2000. (Source: BTM Consult)*[8]

	Total by the end of 2000 (MW)	Added in 2000 (%)
Germany	6107	37
USA	2610	7
Spain	2836	57
Denmark	2341	35
India	1220	18
% of total	82	
Netherlands	473	9
Italy	425	17
UK	424	53
China	352	34
Sweden	265	20
% of total	11	
Rest of the world	1396	54
Total	18,449	32

Realizing the figures in Table 6.3 (spectacular growth concentrated in just a few countries) could lead to the conclusion that, in principle, the present world wind energy market is unstable. If market conditions in one of those countries change, this could have a significant impact on the world market and the manufacturing industry. It is a fact that in individual countries, even those belonging to the top five, the annual market volume varies considerably. These variations can very clearly be seen from the country profiles that are published each year by BTM Consult.[8] The most dramatic changes occurred in the USA, where the annual market volume dropped from 400 MW/yr in 1985 to about 50 MW/yr in 1988. In 1997 the volume increased from about 30 MW/yr to 570 MW/yr in 1998. Also India showed dramatic variations: from 370 MW/yr in 1995 to 80 MW/yr in 1998. Even Denmark, which has shown high growth figures, shows relatively large variations. The only countries that until now appear to have continuous growth are Germany and Spain. The UK has periodic growth and a decrease of volume over a period of about 3.5 years. During the last four years growth in the Netherlands has been low but constant.

There are no simple explanations for these trends. Often changing governmental policies, uncertainties in revenues and planning procedures are the basic reasons why the number of machines installed annually vary so much. Referring to the present successes in Denmark, Germany and Spain, some professional associations claim that the most effective incentive for market development is fixing a minimum feedback tariff for a sufficiently long period, e.g. 10 years. It is evident that in countries like Denmark, Germany and Spain, fixed tariff systems have indeed been very successful. Of course an attractive feedback tariff system is not the only condition for successful development. Additionally a proper national energy or environmental policy with specific renewable energy targets, R&D facilities, an appropriate legal planning framework, public acceptance and an industrial homebase are needed.

In the near future the co-existence of different systems in the open (European) market may become problematic. Introducing a tradable green certificate system, for instance, next to (high level) fixed tariff systems, might lead to a situation where country A, for example, generates the required renewable energy quota and takes all the burdens for country B. Although country B would be buying the certificates it would not be stimulated to develop its own renewable energy sources or its own industry. A careful matching of national systems would therefore be needed for a market with equal opportunities and fair burden sharing.

Nevertheless it would not be too difficult to create market conditions in a large number of countries similar to those in Germany and Spain, provided the wind regimes and infrastructural circumstances allow it. Further expansion of the market in the future would thus be very feasible.

6.2.2.2 Offshore applications

As the best (i.e. wind regime and grid connection) land locations have been used up and public acceptance starts to wane, coastal countries start investigating and exploiting near and offshore resources. Denmark, the Netherlands, Sweden and the UK have already accumulated experience with near-shore wind farms, using existing technologies from the wind turbine and civil engineering industries. See Table 6.4.

In Europe some five near-shore wind farms, varying in capacity from 4 to 150 MW and totalling almost 450 MW, are in the design stage. All new wind farms will have wind turbines with capacities exceeding 1.5 MW. Capacities totalling several thousands of MW appear in the long-term government plans of Denmark, Sweden, Belgium and the Netherlands; Germany, Spain, the UK and others are likely to follow suit. At present all project designs are derivatives of machines designed for land applications, equipped with specific features for operation in water.

Table 6.4 Overview of near-shore demonstration and commercial wind farms, built before 1999

Location	Year of first rotation	Installed power (MW)	Status	Remarks on foundation
Nogersund, Baltic (S)	1991	1 × 0.22 = 0.22	Abandoned in 1998	Tripod on solid rock
Vindeby (DK)	1991	11 × 0.45 = 4.95	In operation	Box caisson on sandy soil
Medemblik, IJsselmeer (NL)	1994	4 × 0.5 = 2	In operation	Steel tower driven in sandy soil and fresh water, corrosion lifetime 50 years
Tunø Knob (DK)	1995	10 × 0.5 = 5	In operation	Box caisson on sandy soil
Dronten, IJsselmeer (NL)	1996	28 × 0.6 = 16.8	In operation	Turbines located a few metres outside dike foot on monopiles in fresh water
Bockstigen, Valar, Baltic (S)	1998	5 × 0.5 = 2.5	In operation	2.25 m steel monopiles secured by grouting in 10 m deep holes drilled in limestone rock

6.2.3 Industry and employment[5,8]

In 2000 the European industry produced more than 90% of total worldwide sales in that year, total turnover of the wind energy sector worldwide was slightly more than €4 billion, growth of turnover was 17% compared with 1998 and total employment, both direct and indirect, of the wind industry was close to 17,000 man-years/yr, which corresponds to about 20 jobs/MW installed power. According to the EWEA/Greenpeace scenario the labour intensity will gradually decrease to a level of about 12.3 man-years/MW by the year 2020.

Of the top manufacturers in terms of sold MW in 1999 three were Danish, four Spanish, two German, one German-USA, one German-Danish and one Japanese. The others produced 5.4% of the total volume. The four largest manufacturers exported between 75% and 93% of their output.

6.2.4 Economy

International organizations without any links to wind power (for example, the International Atomic Energy Agency, IAEA) estimate that in many countries wind power will soon (2005 to 2010) be competitive with fossil and nuclear power when external/social costs are included.

The economics of grid-connected wind power can be evaluated from two different perspectives.

The first is that of public authorities or energy planners making assessments of different energy sources. Here the focus is on levelized cost. These calculations exclude factors determined by society or governments such as inflation and tax.

The second perspective is that of the private or utility investor, where inflation, interest rates, tax, amortization period, etc. are most relevant. Here the focus is on cash flow in each project, on payback time and present value of the investments. Consequently the economics of wind energy differ significantly from country to country.

Here, reference will be made to costs based on an explicit model containing the real relevant physical and economic parameters by means of which deviating economic analyses can be made. The issue of the value of wind-produced electricity (avoiding fossil fuel and environmental cost, capacity credit – to be improved by output forecasting models, etc.) will not be addressed.

The generation cost of wind energy is determined by the following parameters:

- total investment cost, which consists of production, transportation and erection cost of wind turbines
- project preparation costs (permits), cost of land, cost of the infrastructure, etc.
- operation and maintenance cost
- average wind speed at the particular site at hub height
- availability
- technical life time
- amortisation period
- real interest rate.

Ex-works cost has been estimated by surveying commercially available wind turbines belonging to the top 10 of units sold. The survey shows average ex-factory costs of around €350 /m² rotor swept area. Assuming a specific installed power of 450 W/m² swept rotor area this translates into €780 /kW of installed capacity. Note, however, that the generating capacity is determined primarily by its swept rotor area and the local wind speed and not by its installed power. The installed power should be properly matched to the wind speed and rotor swept area in order to achieve an optimum energy output and capacity factor (defined as the energy output divided by the output if the machine were running at rated power during 100% of the time) for the best economy. For this reason manufacturers supply wind turbine types with different values for the specific installed power. The higher this value the more suitable the wind turbine is for higher wind speeds. Actual ex-factory prices in terms of cost per m² swept rotor area may vary plus or minus 20% compared with the key figures mentioned above. Machines with the lowest cost per unit of installed power are thus not necessarily the most economic for a particular site. The above ex-works costs are based on published material such as brochures. The actual price of course depends on negotiations and the number of units acquired during one transaction, e.g. when purchasing machines for wind farms. At present (mid-2000) the number of machines per transaction may amount to as much as 2000 units.

Project preparation cost depends heavily on local circumstances, such as condition of the soil, road conditions, proximity to electrical grid substations, efforts to acquire land and permits and to overcome possible public resistance, etc. In some countries, such as the Netherlands, the procedures to acquire permits are longer than the typical innovation time of a wind turbine type. As a consequence the technology to be applied is outdated the moment permits are granted and the project construction can start. Investigations indicate that project preparation costs in Denmark for wind projects with 600 kW machines on average amount to 1.25 times the ex-factory cost.[7] This does not include grid reinforcement costs and long-distance power transmission lines. Project preparation cost per machine can be reduced considerably by wind farm operation.

Operation and maintenance costs include service, consumables, repair, insurance, administration, possible lease of site, etc. Danish and German experiences indicate that the annual operation and maintenance cost for new category B wind turbines are approximately 0.6 to 1.0 Eurocent/kWh, one-half of which consists of insurance cost. For machines over 10 years of operation these costs may increase to a level of 1.5 to 2 Eurocent/kWh.

Availability, defined as the capability to operate when the wind speed is higher than the wind turbine's cut-in wind speed and lower than its cut-out speed, is typically more than 98% for modern machines.

Technical design lifetime for modern machines is typically 20 years. However, this does not exclude the need to replace certain components after a shorter interval. Because of the rapid development towards large machines uncertainty about the reliability of components of category A machines increases; maintenance costs also tend to increase slightly. This is expected to be a temporary phenomenon. Consumables such as oil for gearboxes, brakes, clutches, etc. are often replaced at intervals of one to three years. Parts of the yaw system are replaced at intervals of five years. Depending on the operational strategy and design, components exposed to fatigue loads such as main bearings, bearings in gearboxes and generators are expected to be replaced halfway through their lifetime. Sometimes the reason for early replacement is design errors, often caused by the need to reduce costs by ignoring safety margins.

Choice of **amortization period** (or economic lifetime) depends on:

- the economic model to be used (levelized cost method or private economy perspective)
- the type of investor and its financial strategy.

From the levelized cost perspective the amortization period is often taken to be equal to technical lifetime. A serious complication is that the innovation speed in the wind energy sector is so high that the technology 'ages' during the project preparation phase.

The **average annual wind speed** on the site is of paramount importance to the cost of energy. As a rule of thumb the annual energy output of modern wind turbines can be estimated by means of the following expression: $e = 3.2 \times V^3 \times A$ (kWh/m² swept rotor area), given that the installed power is properly matched to the rotor swept area and the local wind speed at hub height, and the shape of the frequency distribution curve is close to those found in the coastal areas of Europe. V (m/s) is the annual average wind speed at hub height and A is the rotor swept area in m².[9]

At the best locations in the Netherlands, northern Germany and Denmark annual outputs of over 1000 kWh/m² are often achieved (1000 kWh/m² corresponds to an annual wind speed slightly higher than 5 m/s at 10 m height with average new technology). At UK sites wind speeds of 8, 9 and 10 m/s are common, and annual outputs of over 2000 kWh/m² are being recorded. The average normalized (to a normal wind year) annual output of all wind turbines in Denmark in the period 1996–1998 was 937 kWh/m².[7]

The **cost of wind energy** can be exemplified assuming the following:

ex-works cost (note the slight increase in cost for larger machines, Figure 6.15)	€350 / m²
total investment	125% of ex-factory cost
annual O&M cost	1 Eurocent per kWh
availability	99%
technical lifetime	20 years
amortization period	20 years
real interest rate	5%

The costs were calculated by means of the following expressions:

$$c = \frac{a.I_{tot}}{A.E} + m$$

where:

c = cost (€/kWh)
a = annuity factor
I_{tot} = total investment per m² rotor area
A = availability
E = annual energy output per m² rotor area (kWh/m²)
m = operation and maintenance cost

while:

$$a = \frac{i(1+i)^n}{(1+n)^n - 1}$$

where:

i = real interest
n = amortization period

while: $$E = 3.2V^3 \ (kWh/m^2)^9$$

where:

V = average annual wind speed (m/s) at hub height.

The results are presented in Figure 6.12.

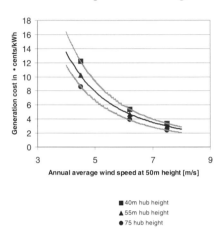

Figure 6.12 Generating cost of wind energy as a function of local average wind speed. The calculations have been carried out for three sizes of wind turbines. The ranges of inaccuracy are due to varying wind speeds at different hub heights. To take into account various dimensions, outputs were calculated at 40 m hub height (typically 500 kW machines), 55 m hub height (typically 1 MW machines) and 75 m (typically 2 MW machines) for a roughness class 2 terrain according to reference 1. A wind speed of 7.5 m/s at 50 m height corresponds to the coastal areas of Ireland, Scotland, northern Denmark and sections in southern France; 6.2 m/s to large areas of Denmark, northern Germany, Netherlands, the UK, Ireland, France, Spain, Greece and some sections of Spain, Portugal and Italy; 4.5 m/s to large parts of the European land mass. For reference see also Figure 6.2.

6.2.5 Non-technical implementation issues

Wind energy's environmental impact has been investigated thoroughly in both Europe and America. The main conclusions are:

- **Atmospheric emissions.** No direct atmospheric emissions are released during the operation of wind turbines. The emissions during production, transport and decommissioning of a wind turbine depend mainly on the type of primary energy used to produce the steel, copper, aluminium, plastics, etc. from which the wind turbine is constructed.
- **Energy balance.** The energy needed to produce, install, operate, maintain and decommission a typical wind turbine is equal to the energy output during a period varying between three and six months. This is fundamentally different from fossil fuel fired plants, which continuously need energy (oil, coal, gas) to produce electricity. In this sense wind turbines, along with solar and biomass plants, are the only net energy producers.
- **Social costs and liabilities after decommissioning.** Electricity from wind turbines has very low external or social costs and no liabilities related to decommissioning of obsolete plants. Almost all parts of a modern wind turbine can be recycled.
- **Land use.** Wind energy density is low and thus harvesting significant amounts of energy requires large areas of land. As a rule of thumb wind farms require

0.06–0.08 km²/MW (12–16 MW/km²). However, the land on which wind turbines are built can still be used for other purposes such as agriculture and cattle farming. Foundations, utility buildings and access roads occupy about 1% of the land.

On the typical, flat, land-based sites wind turbines do not cause erosion and they do not influence vegetation or fauna. In most countries wind power developers are obliged to minimize any disturbance of vegetation during the construction of wind farms and its infrastructure works on sensitive locations such as mountains.

- **Acoustic noise emissions.** Acoustic emissions from wind turbines are composed of a mechanical and an aero-acoustic component, both of which are a function of wind speed. Reducing noise originating from mechanical components is a straightforward engineering exercise, whereas reducing aero-acoustic noise is a rather difficult trial and error process. In modern wind turbines mechanical noise seldom causes problems. Figure 6.13 illustrates the reduction of the acoustic source level of wind turbines in the course of time, resulting from R&D into this area.

  The nuisance caused by turbine noise is one of the important limitations of siting wind turbines close to inhabited areas. The acceptable emission level strongly depends on local regulations. One example is the Dutch regulation for 'silent' areas, where a maximum emission level of 40 dB(A) near residences is allowed, at a wind speed of about 5–7 m/s. At this wind speed level the turbine noise is most distinctly audible. In Europe an acceptable minimum distance between wind turbines and living areas is 150–200 m.

- **Visual impact (landscape, shadow, reflections).** Depending on the characteristics of the landscape (small-scale elements or large-scale industrial or empty plains) modern wind turbines with a hub height of 40 m might have a disturbing visual impact on the landscape, which a number of inhabitants may find difficult to accept. This visual impact, although very difficult to quantify, often poses severe difficulties in the planning process in most European countries. Another form of visual impact is the effect of moving 'shadows' and reflection of sunlight from the rotor blades. This is only a problem in situations where turbines are sited very close to workplaces or dwellings. The effect can easily be predicted and avoided through careful siting. Wind turbines in the MW range can cause shadow effects, which can be more disturbing than acoustic noise effects.

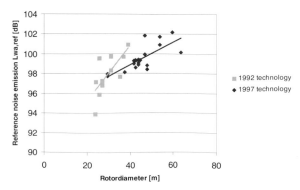

Figure 6.13 The reduction of the acoustic source level of wind turbines in the course of time.[4]

- **Impact on birds.** From studies in Germany, the Netherlands, Denmark and the UK it appears that wind turbines do not pose any substantial threat to birds. The number of birds killed by wind turbines is small compared with birds killed by natural causes, hunting, traffic and high voltage lines. Only isolated examples have been reported, such as the Spanish wind farm of Tarifa near the Strait of Gibraltar, which is located near a major bird migration route. However, in principle disturbance of breeding and resting birds can be a problem on coastal sites.
- **EMI (electromagnetic interference).** The propagation of electromagnetic waves from navigation and telecommunication systems may be influenced by wind turbines. The waves will be scattered and diffracted and as a result interference could occur, which may lead to malfunctioning systems. Investigations by the BBC in Britain and the PTT in the Netherlands, among others, concluded that interference can easily be avoided by proper siting (e.g. keeping a certain distance from the axis of relay connections) and/or the use of antennas with enhanced directivity or gain.
  The IEA has published preparatory information on this subject,[10] identifying the relevant wind turbine parameters (diameter, number and radar cross-section of blades, speed, etc.) and the relevant parameters of the potential vulnerable telecom services (spatial positions of transmitter and receiver, carrier frequency, polarization, etc.).
- **Personnel safety.** Accidents with wind turbines involving people are extremely rare and there is no recorded case of persons hurt by parts of blades or ice loosened from a wind turbine. Insurance companies in the USA, where most of the experience with large wind farms has been accumulated, agree that the wind industry has a good safety profile compared with other energy producing industries. The International Electrical Committee (IEC) has issued an international standard on wind turbine safety.[11]
- **Public acceptance**. This issue has two aspects: the attitude of the general public towards wind energy and the attitude of people living near (planned) wind farms. On the one hand, the general public in many countries favours renewable energy sources such as wind power. On the other hand, deploying a wind farm in a community sometimes raises resistance due to the neighbours' uncertainty and negative expectations about the wind turbine's visual impact and noise emission. This has been called the NIMBY (Not-In-My-Back-Yard) dilemma. Experiences from all over the world indicate that the potential neighbours' participation in planning and ownership of a wind farm can secure the public's acceptance and active support of wind power.
  Opinion surveys in areas of the European Union with many wind turbines or clusters (such as Denmark and the UK) showed that 70–80% of the population is 'generally supportive' or 'unconcerned' with respect to wind turbines in their neighbourhood. In a referendum in a Danish municipality with a very large number of wind turbines, 77% of the votes favoured more wind turbines, though legal procedures fed by resistance by a number of people are very often the main reason for slow progress in implementing wind energy.

6.3 GOALS FOR R&D

6.3.1 Introduction

In Section 6.1 strategic goals were linked to the success factors that are essential for reaching these goals. Here different aspects of the success factors in R&D are linked to the different categories of application (categories A, B, C from Section 6.2) by means of the actions required. The actions actually constitute R&D priorities to futher stimulate the use of wind energy systems. Table 6.5 illustrates the structure and lists the necessary R&D activities. More explanatory information on most of the listed activities will be given in Section 6.3.2.

6.3.2 Cost reduction of wind energy

6.3.2.1 Introduction

Compared with traditional energy sources wind energy at present is competitive at very good sites, even without the compensation for the environmental advantages. For less favoured sites wind energy can compete if CO_2 credits are taken into account. However, to install 100,000 MW of wind power in the EU economically, fully competitive with fossil fuel and nuclear-based power production, requires further cost reduction of wind power.

If the strategic goal for wind power in Europe is realized, Europe will remain the world's largest market for wind turbines. Having such a strong 'home' market is extremely beneficial to the European wind energy industry as a world player, in terms of maintaining and further strengthening its competitiveness.

Since the appearance of the first commercial grid-connected wind turbines (typically 22 kW) at the end of the 1970s, the generating cost of wind electricity has roughly been reduced by a factor of 10. In the beginning, during the transition from prototypes to medium-scale batchwise production, cost reduction was fast, easy and very significant. Later, as the obvious improvements in design were implemented, progress became slower. Between 1981 and 1995 the generating cost in Denmark was reduced from DKK 1.1/kWh to DKK 0.4/kWh.[4] Since the mid 1990s cost reduction again became slower, but continuous. There is still room for a further cost reduction of 15% on the short term and closer to 50% by the year 2020.[8]

Cost reduction of a market product is characterized by the progress factor. This is the relative cost level compared with the previous one, after each doubling of the cumulative production volume. As the Danish wind energy industry has been in business the longest, at a relatively high production level, this industry can be considered as representative of wind energy technology in general. From 1980 to 1995 the progress rate was fairly constant at a level of 80%. Of the 20% cost reduction, about 15% is due to design improvements and more efficient manufacturing and about 5% is due to improved siting. There are no reasons to assume that this ratio will change as the market develops further. Further cost reduction, however, is not an automatic process. Different activities have to be undertaken depending on the elements that determine

Table 6.5 Summary of R&D activities in relation to success factors and applications of wind energy systems. Those in the technology domain (non-shaded) are being addressed explicitly

	Large-scale wind farms on/ offshore: P > 1.5 MW	Distributed systems on land: 0.5 MW < P < 1.5 MW	Decentralized systems: P < 0.5 MW
Cost reduction of wind energy	• Design improvement • Manufacturing technology • Transport and installation techniques • Reliability design for offshore	• Design improvement • Manufacturing technology	• Testing and evaluation prototypes
Increasing the value of wind energy	• Output forecasting methods • Controllability of large wind farms	• Output forecasting methods	
Finding new sites	• Offshore resource assessment • Resource assessment mountainous areas • Resource assessment in cold climates • Design conditions and methods adapted to extreme external conditions	• Resource assessment mountainous areas • Resource assessment in cold climates • Design conditions and methods adapted to extreme external conditions	• Resource assessment mountainous areas • Resource assessment in cold climates • Design conditions and methods adapted to extreme external conditions • Market surveys
System development	• Optimization of power electronic converters • Control strategies (for output and power quality control and to stabilize mechanical constructions) • Grid connection	• Optimization of power electronic converters • Control systems (for output and power quality control) for weak grids	• Energy storage systems • Energy management systems • Standardization and modularization • Updating system design methods and verification by means of experiments
Reduction of uncertainties	• Reliability design methods • More accurate resource assessment methods and output prediction calculation methods • Methods for fine tuning of power curves to local climatical circumstances • Development of fast aerodynamic diagnosis methods	• More accurate resource assessment methods and output prediction calculation methods • Methods for fine tuning of power curves to local climatical circumstances • Development of fast aerodynamic diagnosis methods	• System reliability
Reduction of environmental effects and negative social impacts	• Development of low-noise blades • Monitoring effects on birds' habitats • Minimize visual impact • Develop and test participation models for the public	• Development of low-noise blades • Monitoring effects on birds' habitats • Minimize visual impact • Develop and test participation models for the public	• Design systems with minimum electrical storage by means of batteries
Education and HRD	• Joint international R&D programmes in universities • Develop training schemes for lower and medium level technical skills • Establish specialized professorships at universities • Develop educational material for primary schools	• Joint international R&D programmes in universities • Develop training schemes for lower and medium level technical skills • Establish specialized professorships at universities • Develop educational material for primary schools	• Joint international R&D programmes in universities • Develop training schemes for lower and medium level technical skills
Development of policy and instruments	• Obligatory national targets for wind energy • Co-ordination by EC (European Directive) • Evaluate market stimulation programmes and design more effective instruments • Create open European market	• Obligatory national targets for wind energy • Co-ordination by EC (European Directive) • Evaluate market stimulation programmes and design more effective instruments • Create open European market	• Incorporate technology introduction in national rural development programmes and programmes of multi-national organizations (World Bank, UNDP, etc.)

the cost structure. The cost of establishing a wind energy project is depicted in Figure 6.14. The cost distribution of the main elements is based on a Danish analysis presented in reference 12.

Wind turbine Ex factory 80%	• Design features (concept, loads, materials, component, controls) • Manufacturing (methods, economy of scale) • System efficiency • Overhead (marketing, warranties, management)
Wind farm realisation 15%	• Transport • Installation • Infrastructure (civil works, electrical connection, buildings) • Commissioning
Project development 5%	• Financing • Insurance • Permits and licences • Certification • Bidding • Land acquisition • Siting (wind speed) • PR, participation of public, lobbying

Figure 6.14 Cost elements of wind generated electricity. This cost distribution is based on a 0.6 MW wind turbine operating as a single unit or a small wind farm of less than 8 MW.[12] It has to be taken into account that for large land-based wind farms the main contribution of cost is different. The project development cost will be relatively more expensive. For near and offshore wind farms the cost of civil works will be relatively high. It has to be noted that in this figure only the cost of designing a financial structure was included. The effect of applying different financial strategies to the final energy cost of course can be much more than 5%.

To arrive at the cost per unit of generated electricity, the operation and maintenance (O&M) costs have to be included. Especially for wind turbines located on difficult to access terrain, such as on offshore and mountainous areas, this cost element is very important.

Realizing a cost reduction of 50% in 20 years' time is a combination of addressing all factors mentioned. To set detailed priorities, which have a general value, is not possible. Priorities depend on the concept and on the manufacturing and O&M strategy of a particular manufacturer. But some general priorities can be established on the basis of the following considerations.

6.3.2.2 Technology development: larger wind turbines and improved designs

From the very beginning of modern wind energy technology development, starting in the mid 1970s, we have noticed a gradual and consequent growth of the unit size of commercial machines, from 10 m diameter in the mid 1970s to 80 m at present. For the trend since 1992, see also Table 6.6.

Table 6.6 Approximate average size (kW)
of wind turbines installed each year in Europe

1992	200
1994	300
1996	500
1998	600
1999	700
2000	900

The demand side of the market mainly drives the trend towards larger machines. The most important arguments for larger machines are: utilizing economy of scale, less visual impact on the landscape per unit of installed power, and the expectation that the multi-MW machines are needed for exploiting the offshore potential. The effects of the scale on the relative cost of the machines are illustrated in Figure 6.15.

At present the specific weights of the MW class machines are already approaching those of machines in the 0.75–1 MW class: 25.1 kg/m² swept rotor area and 23.8 kg/m², respectively.

Ex-factory cost reductions of 15–20% are being expected from the combination of the following features in advanced wind turbine concepts:[14,15]

- reduction of loads by means of less conservative design and the use of flexible turbine components, such as flexible blades, flexible hubs, variable speed generator systems; this leads to lower weights and lower machine cost
- reduction of the number of components, e.g.: the application of direct-drive generators, passive blade pitch control (in combination with variable-speed drive trains), passive yaw combined with a downwind located rotor
- use of improved materials featuring higher strength to mass ratios and better internal damping. R&D on these aspects has been carried out since the early 1980s and even earlier. The claims of this work are still valid, although there are no success stories in the commercial sense yet. The main reason is that the market pressure is so large that time is lacking for long-term, high-risk developments. In the present

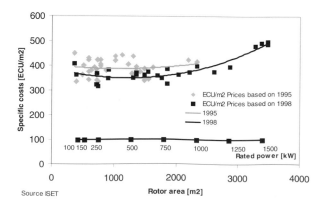

Source ISET

Figure 6.15 Relative wind turbine cost as a function of size in the course of time (ISET 1999).[13]

commercial machines, elements of these innovative design features can already be noticed. An increasing number of manufacturers are offering variable speed conversion systems, although the first ones were offered on the market in the late 1970s (in Europe, Lagerwey later followed by ENERCON, and still later in the USA by former KENETECH). A number of manufacturers (ENERCON (D), Lagerwey the Windmaster (NL)) are offering or developing (Jeumont (F) and ScanWind (S/N)) direct-drive generators with solid-state energy converters. Because of the low content of harmonic distortion, the absence of inrush currents and adjustable reactive power supply, these modern inverters make wind turbines particularly suitable for weaker grids. The market share of wind turbines equipped with direct-drive generators and/or variable speed systems at the end of 1997 was slightly over 30% and is increasing slowly. Figure 6.16 shows two examples of wind turbines equipped with direct drive generator systems.

Measurements of some experimental rotor concepts with different flexible blade attachments in the Netherlands, the USA, the UK, Germany and Denmark have shown a dramatic reduction of dynamic loads on the blades. In classic designs these loads are the most important design drivers of the blades, hub and tower structures.

The promising results of this and other research projects have never been fully implemented in new designs, but as wind turbines become larger and thus intrinsically more flexible, manufacturers may have to reassess these findings and implement the results.

As offshore applications will be very important in the near future in north and north-west Europe, some special remarks have to be made concerning typical offshore design features.

Integration of wind and wave loading into the structural dynamics codes is necessary to avoid overdesign. From calculations it appears that the combined wind and wave-

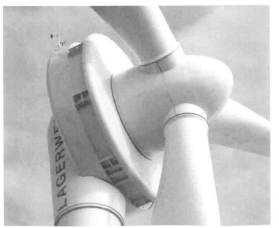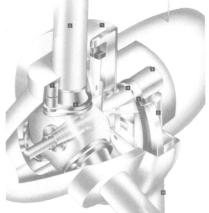

Figure 6.16 Examples of wind turbines equipped with direct-drive generator systems. (Lagerwey the Windmaster, ENERCON)

loading spectrum is lower than the wave loading treated separately. Applying this finding in the design will lead to lighter structures, translating the absence of noise emission constraints (tip speeds of rotors can be higher, leading to smaller transmission ratios and lighter drive trains) and lower turbulence intensities into relatively cheap constructions. Jamieson[16] even claims that these advantages could make up for the inherent costs of offshore conditions.

The final goal of all cost-reducing R&D efforts is to reduce wind electricity generation cost by 15% in the short term (2005) and close to 50% by the year 2020.

6.3.3 Increasing the value of wind energy

As the difference between value and cost increases so the economic drive to realize wind energy plants will grow. For society wind energy, like other renewable energy technologies, represents the following values:

- displacement of fossil fuels, which cause hazardous gas emissions
- displacement of installed power in fossil fuelled and nuclear energy plants
- saving natural resources
- creating sustainable employment
- independence from politically sensitive fossil fuel resources (increased supply security).

It is not clear how these advantages are to be reflected in the economic process. However, R&D should concentrate on the development of methods to quantify and improve the value. The first priorities are developing energy output forecasting models (see Section 6.3.5), improving methods to control power output of wind farms in order to maximize capacity credit, and the replacing of fossil fuels. This becomes particularly interesting when a spot market for renewable energy develops.

The R&D goal is to have a fully quantified value system of wind electricity by the year 2005.

6.3.4 Finding new sites

Already in densely populated countries with a high level of installed wind capacity, the best sites are being occupied. Combined with the complicated physical planning procedures and increasing public resistance against new wind farms, this leads to the need for alternative sites. Nevertheless, the relative average output of wind turbines, based on a large number of machines, is increasing. See e.g. Figure 6.17. This increase is caused mainly by a combination of the use of sites with higher wind speeds and the use of higher towers, but also, although to a lesser extent, by improved system efficiency and availability. The improvement of the overall system efficiency is due to improved aerodynamic efficiency of the rotor blades, the application of high-efficiency electric conversion systems and better matching of the wind turbine rating to the local wind

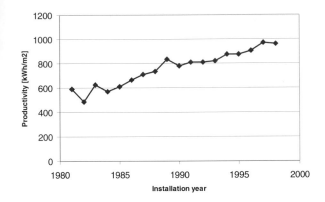

Figure 6.17 Development of the energy output per m² swept rotor area of all Danish wind turbines.[7]

regime. (The latter means that the installed power per unit of swept rotor area has to be in accordance with the local wind speed. To this end, manufacturers offer wind turbines with various generator capacities and one rotor diameter, or wind turbines with various rotor dimensions in conjunction with the same generator capacity.) The overall system efficiency of present commercial wind turbines is close to what is theoretically feasible.

As developments proceed and the best sites are being utilized, the future might bring a gradual decrease of the output curve (Figure 6.17), indicating the saturation of wind potential on land. This effect can be delayed by replacing the older and smaller wind turbines by state-of-the-art large units. Countries with a coastline will have the opportunity to continue the exploitation of wind potential by using offshore sites. See Section 6.2 for initiatives in various European countries.

In addition, non-conventional inland locations might be considered. In Germany 45% of the total installed power is realized in inland states (*Länder*). In many cases extended towers are being used to compensate for the reduced wind speeds at 'normal' heights.

Other extreme locations may include mountainous sites with extreme high wind speeds and elevated turbulence intensities.

In Europe Finland, Sweden and Norway, among others, are deploying wind turbines in extremely cold climates.

Dedicated wind turbine types will be developed for low wind speed inland locations, high wind speed–high turbulence locations, cold climates and offshore locations.

The specific R&D goal is to complement the European Wind Atlas *with measured data of the North Sea and a number of high potential mountainous sites.*

6.3.5 System development

European utilities consider wind power more and more as a viable and reliable source of energy in the grid's supply portfolio. However, there is still some reluctance, especially in inexperienced countries, because of the utilities' concern about power quality, power predictability and grid integration. Furthermore, Europe's electrical utility sector is facing a significant restructuring process (generation, transport and distribution are being separated, whilst decentralized generation is now the main responsibility of the distribution sector), leaving less attention to the introduction of wind power and other renewable energy sources.

As the widespread use of wind power in Europe demands the active participation of European utilities, wind power's credibility from the perspective of the utilities must be increased. The most important means to this end are:

• ensuring wind energy's power quality (harmonic distortion, reactive currents, inrush currents, voltage stability) and predictability (see Section 6.3.3)
• reducting wind power's transmission costs and grid costs
• maximising the quality of the electricity produced.

The R&D goals are to develop electrical output prediction tools that predict output 24 hours in advance for wind farms with a standard deviation of 5–10% by 2010, and to arrive at technical and economic optimal planning tools for integration of 100,000 MW of wind power into the European grid by 2010.

Virtually all installed electricity-generating wind turbines are grid-connected machines. Often the exploitable wind potential is high in areas with a poor electricity infrastructure. Modern power electronics make wind turbines more suitable to connect to weak grids, as addressed earlier.

A very important application – electricity supply systems to remote areas without an electricity grid – never developed commercially on a significant scale. Wind–diesel systems providing electricity to small to medium sized autonomous grids were tested and demonstrated on a limited scale.

Apparently the need for autonomous systems has not changed since the 1980s. However, the technology has improved considerably. Those components that in the past appeared to be too expensive or too unreliable, such as wind turbines, electronic components and control strategies, have improved dramatically since. This leads to the conclusion that a revival of the development of autonomous systems is very possible. The first signs are already visible, e.g.: ENERCON (Germany) started the production of a 0-series of stand-alone desalination units; Atlantic Orient Company (USA) installed 11 wind–diesel systems, comprising 35 wind turbines, in extreme climates (Alaska, Siberia); Lagerwey the Windmaster (Netherlands) is developing new systems; Northern Power Systems (USA) supplies a number of specialized autonomous and stand-alone units including wind turbines and PV arrays.

The standardization of subsystems to increase the modularity of systems is a necessary condition for accelerated market development. It will lead to reductions not only in the production cost of systems, but also in the design and engineering costs.

The R&D goal is to have co-ordinated system development and testing programmes in place in all major European R&D centres, in which the industry, familiar with the market for autonomous energy systems (including battery chargers and wind pumps), is fully involved. The first full testing programme should be operational by 2005.

6.3.6 Reduction of uncertainties

Finance and insurance companies consider wind power as a reliable source of energy and have confidence in the overall performance of the wind energy industry. Within the overall necessity to utilize all possibilities to optimize cost, services provided by the financial and insurance companies will become cheaper if uncertainties in assessing both technical and economic risks can be decreased. Most significantly this applies to determining wind resources, the prediction of the wind turbine's performance, and prediction of O&M cost.

6.3.6.1 Wind resources

Earlier the importance of the wind regime for the economic performance of a wind energy project was addressed (see Sections 6.2.4 and 6.1.2). Additionally, non-surveyed terrains with an obvious high potential should be mapped and added to the *European Wind Atlas* project, which is specifically concerned with detailed measurements of near and offshore areas and mountainous regions.

6.3.6.2 Wind turbine performance prediction

There is still considerable room for improvement in the present level of accuracy. It is a basic problem of stochastic data and non-linear behaviour of the systems that poses considerable theoretical problems. Necessary activities include improvement of anemometry, terrain calibration, and transfer methods of measured power–wind speed curves of wind turbines from one site to the other.

It often appears that the actual performance of wind turbines differs considerably from the predicted one, depending on the local climate or season, leading to under- or overproduction of energy (overproduction often leads to accelerated reduction of a turbine's life and is not acceptable). In those cases manufacturers may be faced with financial claims from the project owners. One solution is to have methods for fine tuning the aerodynamic performance of rotor blades and have available associated fast field diagnosis instruments.

A particular aspect, mentioned earlier, applies specially to large wind turbines. The actual dynamic behaviour regularly differs from analyses carried out during the design stage. If prohibitive vibrations appear during operation, involved costs do not permit the wind turbine to be replaced by another one. In those cases control systems can be used to actively damp vibrations after the turbine is put into operation. See, for instance, reference 17.

6.3.6.3 Operation and maintenance

Figure 6.10 showed how availability (and thus energy output) depends on reliability and accessibility of the wind turbine installations. Wind turbines located in offshore locations and in mountainous terrain are especially subject to potential very high cost due to increased O&M cost and loss of availability. In this respect the accelerated development of early failure detection and condition systems is essential. These systems are to be connected to planning systems for maintenance. Data may be used as inputs for improving the reliability of the wind turbine design.

The condensed R&D goal for the various activities described here can be formulated as follows: the uncertainty of generation cost for flat terrain and offshore should be reduced to 5–10% by 2010. For complex (mountainous) terrain uncertainty should be reduced to 10–15% by 2010.

6.3.7 Reduction of environmental effects

If 100,000 MW of wind power is to be installed in Europe over the next 20 years it is essential to maintain and increase the public's acceptance of wind power. R&D efforts to minimize the environmental impacts and to increase the public's involvement, e.g. by developing participation models, should be balanced with those needed for the technical aspects.

R&D goals include the availability of European standards and codes of practice for noise emissions, siting, physical planning, minimizing impact on birds' habitats, etc.

Some of these standards and codes of practice have to be based upon R&D results, such as design methods for low-noise blades and monitoring of birds' behaviour near wind farms in sensitive areas.

6.3.8 Education and human resource development

A workforce, skilled in both technical and non-technical aspects of the technology, is an absolute necessity to meet the strategic goals. It is the responsibility of the national ministries responsible for education and science to incorporate wind energy technology in the different school programmes on different levels. These efforts need to be closely co-ordinated with similar activities concerning other renewable technologies.

The specific goal is to have at least one European, formally accepted, university degree on renewable energy, with a specialization option on wind energy technology.

6.3.9 Policy and instruments

The first three activities listed under 'Development of policy and instruments' in Table 6.5 do not need explanation. The issue of creating an open European market is less obvious.

As mentioned in the European Commission's Green Paper *For a European Union Energy Policy*, the free operation of the market has to be the principal instrument of any policy for wind energy. Also non-EU-based actors must have access to this market.

One important point is to develop harmonized European standards, legal structures and institutional frameworks on the essential aspects of market operations.

In this respect the following goals are to be met:

* *European planning procedures for ranking feasible sites (on flat coastal and inland sites, in mountains and offshore) for 10,000 MW by 2000 and 100,000 MW by 2010*

- *a European certification and accreditation system facilitating a free market for wind power technology by 2005*
- *a European standard for performance (test and measurement) by 2000*
- *a European standard for risk assessment for investment in renewable energy by 2000*
- *a transparent, harmonized cost and tariff system facilitating a free market for wind power by 2005*
- *a European code of practice to connect wind power to the grid system and the reinforcing of the grid to enable grid integration.*

6.4 ROADS TO MARKET

6.4.1 Land-based wind energy systems in industrialized countries

Although the share of wind energy in annual, newly built electricity-generating capacity is still small (in 1999 10% in terms of MW installed, which means approximately 3% in replaced capacity), wind energy in principle has developed into a common option in the planning of electricity generating capacity.

In industrialized countries with suitable wind regimes and adequate infrastructure, wind turbine systems do not need to be demonstrated on land locations.

Finance is not a major problem either as financial institutions have gained sufficient confidence in wind energy technology. Thus (initial) investment support is not needed.

Some countries have a high population density, and conflicts often arise in the quest for land on which to build wind farms. Therefore, problems related to conflicting land use need to be resolved by means of *policy development, acquiring broad public support* and the *development of public participation models* for which the liberalized electricity market provides a very good base supported by a variety of *legal measures that support national targets*.

Naturally this requires *firm national targets per renewable energy technology band*. These targets need to be established in national energy and environmental policies. For the members of the European Union, the European Commission could be a strong catalyser, which could enforce the implementation of national targets, check progress and implement a fair means of burden sharing between member states.

Economic conditions need to be created so that the industry can operate on a level playing field, meaning that barriers and subsidies for fossil-fuel-based electricity generation should be removed or, alternatively, feedback tariffs should be fixed for a sufficiently long period to compensate for the external advantages of wind energy. Minimizing the cost of wind turbine systems is a task of the industry (they are forced to because of competition), but maximizing the value of wind energy is the responsibility of national governments and the European Union.

The role of R&D in following the road to market is to identify problems and try to solve them. This can be achieved by developing and evaluating participation models and policy instruments, reducing acoustic noise emission, gradually extending or modifying the existing distribution grid so it can absorb a large amount of electricity generated at a large number of locations distributed over a large area, etc.

6.4.2 Land-based systems in developing economies

All recommendations formulated in the previous section apply in principle to these countries. However, the impact of the general public's involvement is much less. On the other hand, the electric infrastructure in most cases is less well developed and not designed for the fast growing demand for electricity. The fact that the grid is unreliable poses specific problems for grid-connected wind turbines. *Design specifications from the point of view of electrical system and control design have to be derived from these circumstances.*

Wind energy can contribute to these economies in terms of creating local manufacturing capacity and associated employment. In this respect transfer of know-how is essential.

The R&D communities, both in the industrialized North and in the developing world, are to play a vital role in system design, know-how transfer and the building-up of local expertise and human resources.

6.4.3 Wind energy systems for remote areas

Earlier it was concluded that the technology of autonomous (hybrid) systems is insufficiently developed and therefore the market does not develop. Because of this the industry finds the risk of investing in the development of autonomous systems and its potential markets (distribution and after-sales systems, expertise) too high. As most of the markets concerned are lacking financial means this vicious circle is very hard to break. *The first priority has to be given to market development*, before dedicated R&D efforts can be proven. As the demand side of the market and financiers are not sufficiently familiar with the technology and thus confidence is lacking, the proper way of initializing the market (again) seems to be *large-scale demonstration programmes and investment support schemes after potential markets have been identified*. In parallel, technology-oriented R&D programmes carried out by the industry (preferably that part of the industry familiar with this typical market segment) in co-operation with R&D institutes should be (re-)initiated.

6.4.4 Offshore applications

The applications fall into a gradual spectrum of offshore conditions: from near-shore wind farms in shallow water to far offshore, deep-water floating farms. The differences between each type are water depth, distance to the shore, maximum wind speeds and wave heights. Apart from the impact on the design specifications for the machines, transport and installation equipment, these external conditions are the reason for a major difference in the accessibility of the wind turbine systems and therefore operation and maintenance strategies and cost.

Experiences so far (see Table 6.4) do not suggest implementation strategies for nearshore applications that are different from those for land-based farms. All technologies applied are adopted existing technologies from the wind turbine and civil engineering sector. *Monitoring of environmental impacts is essential in order to provide information for future near and offshore projects. Near-shore projects are being projected in environmentally sensitive areas.*

The situation for offshore applications is basically different, as has been pointed out in the previous chapter. Here a number of first-of-a-kind prototype projects seem appropriate, where a limited number of newly developed wind turbine concepts in combination with the support structure, transport and installation methods and O&M strategies can be tested. The European Commission could play a key role in initiating, co-financing and co-ordinating these rather large-scale projects. Even if the big multinational energy companies are taking their own initiatives, a multinational governmental body, such as the EC, is essential as many legal uncertainties, especially in the non-exclusive economic zones, are likely to be encountered during the planning process.

In both cases the role of the wind turbine industry, the offshore industry, contractors, national governments and the R&D community is essential as far as technology development, assessing environmental impacts and the development of legal structures (primarily to control possible conflicts of interest between the various users of the open sea: defence, fishery, shipping, oil and gas exploration and mining, sand mining, nature values, etc.) are concerned.

6.4.5 General

As the demand for wind energy systems developed very fast, manufacturers were not in the position to fully incorporate R&D results in their designs in order to optimize the systems. This implies that lessons from other technologies have not yet been learned. The full use of R&D results is needed in order to be able to utilize the cost reduction potential to its full extent. Reducing R&D budgets at the very moment that wind energy technology appears to become successful in the market place is therefore very short-sighted.

On the contrary, R&D efforts should be intensified, but targeted on those problems that are likely to hamper the large-scale implementation speed of wind energy in the future.

REFERENCES

1 I Troen and E Petersen, *The European Wind Atlas*, Risø National Laboratory, Roskilde, Denmark, 1989.
2 J Beurskens, 'Possibilities for offshore applications of wind turbine systems in Europe', *Proceedings Husum '99, Messe und Fachkongress für Windenergie*, Husum, 22–26 September 1999, 2000.
3 L Germanischer, *Study of Offshore Wind Energy in the EC*. JOULE I project, Verlag Natürliche Energie, Brekendorf, Germany, 1995.
4 P Dannemand and P Fuglsang, *Vurdering af udviklingsforløb for vindkraftteknologien*. Risø National Laboratory, Roskilde, Denmark, 1996.
5 European Wind Energy Association (Forum for Energy and Development) and Greenpeace

International, *Wind Force Ten. A blueprint to achieve 10% of the world's electricity from wind power by 2020*, EWEA, Brussels, 2000.

6   M Kühn, M C Ferguson, B Göransson, T T Cockerill, R Harrison, L A Harland, J H Vugts and R Wiecherink, Opti-OWECS; *Structural and Economic Optimisation of Bottom Mounted Offshore Wind Energy Converters*. JOULE III project, Delft University of Technology, Netherlands, August 1998.

7   Energistirelsen; Miljø- og Energiministeriet, *Wind Power in Denmark: Technology. Policies and Results*. Risø, November 1998.

8   BMT Consult, *International Wind Energy Development: Market Update 1999; Forecast 2000–2004*. Ringkøbing, Denmark, March 2000.

9   J Beurskens, P Andersen, E L Petersen and A Garrad, 'Wind energy', *Eurosun '96 Conference*, Freiburg, Germany, 16–19 September 1996.

10 IEA Wind Energy Programme, *Recommended Practices for Wind Turbine Testing and Evaluation: Electromagnetic Interference*, Lyngby, Denmark, February.

11 IEC 6 1400-1 Ed. 2, *Wind turbine generator systems. Part 1: Safety requirements*, February 1999.

12 IEA Wind Energy Annual Report 1999, NREL BR-500-27988, Golden (CO), May 2000.

13 M Durstewitz, C Ensslin, B Hahn and M Hoppe-Kilpper, *Jahresauswertung 1997 des WMEP*. Institut für Solare Energieversorgungstechnik ISET, Kassel, Germany, 1998.

14 F Hagg, P Joosse, G van Kuik, H Beurskens and J Dekker, 'The results of the Dutch FLEXHAT Programme: the technology for the next generation of wind turbines', *Proceedings AWEA Wind Power '93 Wind Energy Conference*, San Francisco, 12–16 July 1993.

15 F Rasmussen and J T Petersen, A soft rotor concept: design, verification and potentials', *Proceedings European Wind Energy Conference*, Nice, March 1999.

16 P Jamieson and D C Quarton, 'Technology development for offshore', *Proceedings European Wind Energy Conference EWEC 99*, Nice, March 1999.

17 T G van Engenel, E L van Hooft and P Schaak, 'Development of wind turbine control algorithms for industrial use', *European Wind Energy Conference EWEC 2001*, Copenhagen, September 2001.

BIBLIOGRAPHY

European Commission, European Wind Energy Association, *Wind Energy: The Facts*. Altener Publications, Luxembourg, 1999.

N G Mortensen, D N Heathfield, L Landberg, O Rathmann, I Troen, and E L Petersen, *Wind Atlas Analysis and Application Program: WasP 7.0 Help Facility*. Risø National Laboratory, Roskilde, Denmark, 2000.

K Rehfeldt, 'Windenergienutzung in der Bundesrepublik Deutschland; Stand 31-12-1999', *DEWI Magazin*, Nr 16, Deutsches Windenergie-Institut, Wilhelmshaven, Germany, February 2000.

7. INTEGRATION OF RENEWABLE ENERGY SOURCES INTO ENERGY SYSTEMS

A Zervos, D Diakoulaki (NTUA), D Mayer (ARMINES)

7.1 INTRODUCTION AND STRATEGIC SUMMARY

The EU's target of doubling the current share of renewables by 2010, thus attaining a 12% contribution to the EU's gross inland consumption, is generally agreed to be a realistic goal and a necessary first step in complying with the international commitments of environmental protection.

Despite the remarkable technological progress achieved recently and the increasing competitiveness of renewable energy (RE) technologies, the large-scale integration of renewable energies into the European energy system is not straightforward. Available technical options are capable of satisfying ambitious targets only if considered from an integral planning perspective and used in combined schemes aimed at maximizing efficiency. This means that significant changes have to be made in existing energy infrastructures for accommodating renewable energy systems (RES) in an optimized way. Moreover, if in the long term renewables are intended to play a prominent role in the total energy supply a radical renewal of energy infrastructures has to take place.

In concept RE systems and schemes can be envisaged to consist only of RE technologies for up to 100% local power supply from centralized RE systems and integration of dispersed RE schemes into the existing power supply. In practice the power supply will often include fossil fuel energy sources for support and backup, in particular when RE systems are introduced in supply systems with existing fossil fuel power plants. Besides, RE technologies, when used on their own, have shortcomings when compared with the ability of traditional fossil fuel generation technologies to supply firm power with accepted high capacity value. Therefore, in order to utilize the full potential of RE technologies they may be used together in an integrated way comprising hybrid schemes. Such integrated systems could consist of RE generating technologies in combination with other RE technologies and energy storage and/or power conditioning technologies. They can be used in different layouts (centralized systems for local power supply or dispersed schemes for regional power supply) adapted to the conditions and possibilities of each specific location, to ensure firm and reliable power. Ultimately RE systems could develop into completely RE-based schemes that include new types of energy carriers such as hydrogen in combination with new ways to supply and manage the end user's energy demand.

During the 1990s it had been widely recognized that technical progress was a necessary but not sufficient condition for the large-scale integration of RES. Non-technical parameters are playing a major role in the process of their market penetration and in most cases they are supposed to be the most important barriers for RES development. They are either of a political and legislative nature or relate to institutional, environmental and socio-economic issues. All these obstacles are closely related to each other and as a whole reflect a general resistance to change, but also a limited knowledge about technical, economic, social and environmental advantages of RES. It is of primary importance to identify the main sources of resistance in each case and to promote measures aimed at establishing a favourable environment for the large-scale implementation of RES.

With regard to the cost of renewable technologies, existing market imperfections still limit the establishment of fair competition with conventional systems. This failure may severely threaten the further development of RE-generated electricity in the emerging internal electricity market, and therefore undermines compliance with the targets set in the White Paper on RES and the Kyoto commitments. The EU directive 'on the promotion of electricity from renewable energy sources in the internal electricity market' provides a sound framework for minimizing relevant risks.

Furthermore, recent experiences in different EU countries or regions clearly demonstrate that there are ways to successfully cope with these barriers, even in a rapidly liberalizing market. In addition, the new constraints regarding environmental protection and sustainable development establish a very positive framework favouring the option of RES against conventional energy sources. Apart from the reconsideration and specification of financial incentives and pricing policies it is necessary to mobilize the necessary funds through the promotion of new funding mechanisms.

The above remarks make clear that non-technical issues merit particular attention in order to establish and realize plans for RES integration. The main parameters influencing the penetration of RES, in both a negative and a positive way, should be determined and analysed in order to identify the measures to be promoted for removing existing barriers and to take maximum advantage of any favourable conditions. Such a procedure is fundamental, especially in cases where a 100% RES supply is intended.

7.2 STRATEGIC CONSIDERATIONS

7.2.1 Policy and politics

Overall guidelines and priorities should reflect a policy of actively promoting and supporting the development towards a high degree of integration of RE into the European power supply. This should be done in such a way that inter-European co-operation is supported without distorting the competitive balance between the parties in the European resource base.

On a political and strategic level those preferences and initiatives that are needed to implement this policy should be supported or implemented. It could include support for strategic work such as a complete reasoning and argumentation for RE technology (White Paper) in both a regional (EU) and global (worldwide) context, together with implementation and incentive initiatives for market developments. This initiative could also include a policy for the utilization and transfer to the second and third worlds of the European capabilities of technology and experience that have been, and will be, acquired through the European research, development and demonstration (RD&D) efforts.

The policy should emphasize a reinforced, integrated and interdisciplinary approach, including the continuation and formation of interdisciplinary networks. In addition, it should include preference for projects of the 'co-ordinated programme' type, i.e. projects with the potential to cover the entire range from prefeasibility study, feasibility study and demonstration phase (mainly programme financed), to large-scale implementation with (mainly) commercial international financing, and also in countries outside the EU.

7.2.2 Priorities and schedules

Each separate RE technology is presently developing at a rapid and steady pace, and the associated needs for RD&D are dealt with in the other chapters.

In the context of RE integration the needs are primarily in the motivation of the market forces to take advantage of the available technical options and to better accommodate RE systems in existing or new infrastructures. The establishment of fair and transparent competition rules is a prerequisite to increasing the investors' confidence. In addition, a number of 'demand side' issues should have high priority, such as harmonizing standards and increasing technical and economical credibility of the RE technologies. In this context, the development of tools is necessary for planning and assessment of systems and schemes, and for establishing agreed rates and criteria for their evaluation, taking into account environmental externalities and other social costs/values.

The equally important demonstration needs are primarily for carefully designed, well-executed and closely monitored demonstrations of systems and schemes for local and regional integration of RE technologies. Competent and critical review of the systems developed and demonstrated is a precondition for a qualified feedback to the R&D programmes both for integrated systems and for specific RE technologies. In the long term development programmes should be implemented to improve tools and other capabilities, and demonstration programmes should be implemented to show improved concepts and/or applications as they become available.

7.2.3 Key players

Typically, a large variety of actors should be involved in the development and implementation process of RE integration: manufacturers, utilities and energy professionals as well as municipalities, regions and national authorities. Main contractors on the development part should typically be major (national) institutes, while main contractors on the demonstration part typically should be industry based (manufacturers) or end users (communities, utilities).

As long as RE technologies are continuously upgrading and important integration issues still remain unresolved, an equally important role is assigned to research institutions and universities. Moreover, the pressing needs for authorization, standardization and certification procedures call for the creation of competent bodies whose aim is to facilitate potential investors and safeguard the consumers' interests. Finally, the development of the internal electricity market brings out new market players such as regulators and system operators.

Cross-national, cross-discipline and cross-sector co-operation should be facilitated as an important part of building European capabilities, and operational projects should also be a high priority. It would seem preferable to implement several manageable projects, even if they appear to overlap each other, rather than implement a few large 'white elephant' projects encompassing everything and everybody.

7.3   PRESENT SITUATION

7.3.1   RE technologies in Europe

There is presently a significant increase in the interest shown by governments, planners, utilities and private investors in including RE technologies in the energy supply portfolio. A long-term goal of European policy is a significant contribution (12% at year 2010) from European-based renewable energy. This goal is influenced by rising concern regarding externality costs (CO_2 related and others).

Today's most promising RE technologies are based on wind, biomass and solar energy for electricity production as well as for heat production. Currently, RE applications deal mostly with one RE technology at a time such as wind parks, PV systems, etc.

To date international R&D in hybrid RE systems has mostly concentrated on wind–diesel systems. It is envisaged that future R&D programmes will concentrate more on RE systems in which several RE technologies are integrated.

National capabilities exist in manufacturing, consultancy, application and R&D for RE technologies and increasingly in hybrid technologies. European networks and alliances in the field of RE technologies are already established, and more are being formed, very much as a result of EU programmes emphasizing this type of development.

Within each of the RE technologies, a lead position has been established at both national and European levels within manufacturing, consultancy, construction, implementation and application, as well as in research and development.

The R&D effort in the field of RE is in accordance with national and European policies on energy, and the EU Framework programmes have made available considerable allocations of funding for developing RE technologies. Complementary national R&D programmes also exist, with emphasis on different RE topics, according to national resources and preferences.

RE technologies are not yet fully competitive with existing energy-supply technologies on purely commercial considerations, i.e. when externalities are not included in the comparison. However, it is generally anticipated that the new RE technologies (wind, PV, solar thermal, biomass, etc.) will be increasingly economically competitive with fossil and nuclear plant, and some, such as wind, fully competitive within a time horizon of 10 to 20 years. In high-wind locations, wind is already competitive.

European manufacturers, consultants and research institutions are in a good position to contribute very substantially to the enormous effort necessary worldwide during the coming years in RD&D and implementation of RE technologies.

7.3.2 Integrated RE systems

There is a growing realization that RE technologies should be applied in combination with each other, to supplement each other and to improve capacity values.

Today, the situation is that few manufacturers aim specifically at systems, but there are suppliers of hybrid wind–diesel systems (i.e. systems with wind and diesel generation together with another RE technology, typically PV) with development strategies that aim at integrated RE systems. Although there is no widespread application of integrated RE systems, a number of pilot plants and demonstration projects have been implemented. Their operation adds to the increasing body of experience being accumulated.

The development and application of integrated RE systems is presently aimed at the supply of electricity to grids, heat for use in (district) heating and as process heat, and energy carriers such as biogas and biofuels (and in the long run hydrogen from electrolysis).

The integration of renewable energy in the urban environment using passive and active systems has also been considered during the last few years. A number of pilot projects for modern urbanization ('solar city' concepts), which take into account new climatic techniques in order to develop a new solar and bio-climatic architecture, have been designed.

Attention has also been given to the possibilities of producing water using integrated RE systems specifically in regions like the Mediterranean where the natural water supply is limited. A new way of producing water in a sustainable and affordable manner is currently being investigated.

There is a simultaneous development of planning tools in terms of system models on various levels, predicting resources and output, but practically no standards or agreed evaluation criteria are established.

7.4 INTEGRATION ISSUES

In order to promote the integration of RE and increase their penetration a certain number of technical and non-technical issues have to be addressed.

7.4.1 Technical issues

At present there are a number of technical options that can be used for the exploitation of RES according to the specific situation in each region and the availability of local resources. However, for increasing technical efficiencies and for maximizing the degree of RES penetration new effective solutions have to be promoted. These basically consist of using RE generating technologies in combination with other technologies, including energy management and energy storage and/or power conditioning technologies. In addition, the massive RES utilization in existing infrastructures may have unwanted impacts, and layouts should be adapted to the conditions of each specific location to ensure guaranteed and reliable supply.

7.4.1.1 Classification of integrated RE systems
In order for the full potential of RE technologies to be utilized, they must be used together in an integrated way. It is obvious that there exist an immense variety of areas differing in a large number of parameters such as size, population density, climatic conditions, building styles, cultural patterns, resource availability and, of course, energy system characteristics. However, what is of particular interest when examining the prospects for RES integration in an area can be reflected in a small number of characteristics:

- energy consumption density per area unit, compared with RE availability
- availability and kind of energy infrastructure
- power consumption pattern
- size.

Different combinations of RE technologies can be used according to the type and size of the energy system to which they are to be integrated. We can classify them as integrated RE systems under the following headings:

SINGLE CONSUMERS AND SMALL GROUPS
Photovoltaics have proven their success in supplying small and very small consumers. Solar home systems consisting of PV modules, batteries and charge controllers up to approximately 100 W have become one of the most well-established applications. In cases where more power is required, wind energy converters and biomass plants often become more attractive due to their economic performance. In the case of grid absence a storage system and/or a conventional generation backup is needed. Solar

thermal collectors, biomass systems and geothermal heat pumps are well-established technologies, which could cover the heating and cooling needs of consumers.

STAND-ALONE AND ISOLATED GRIDS

In general the decentralized electrification of local communal and regional structures by erecting or expanding stand-alone grids presents a very important potential for the large-scale deployment of renewable energies worldwide. Solar applications, wind energy converters and biomass plants are suggested for medium loads and local power supply. The weak or autonomous grid requires storage systems in order to approve the system reliability. Once again, solar thermal collectors, biomass systems and geothermal heat pumps could cover the heating and cooling needs of consumers.

LOCAL ENERGY SUPPLY

In this category we can differentiate between rural and urban areas. The differentiation is based on the comparison of the energy consumption density with RES availability. In urban areas the RES input is much smaller than the consumption density. The main RE source is the sun, with a limited availability of the other RE resources. In the case of rural areas the RES input is in the range of the energy consumption density. Usually there is a significant availability of several RE resources (solar, wind, biomass, hydro).

REGIONAL ENERGY SUPPLY

Large-scale applications of all the available technologies.
Selection among possible alternatives depends on the following parameters:

• social acceptance
• technical reliability
• sources availability
• economic effectiveness
• environmental protection.

The objectives are:

• to cover the demand in a sustainable way
• to exploit the available renewable energy sources
• to use the most mature and cost-effective RES technologies.

Additional objectives for electricity production are:

• to cover the maximum average net hourly production
• to provide the electrical system with an adequate safety margin.

7.4.1.2   Levels of integration

The level of RES penetration is highly influenced by the type and degree of adaptation of the current energy infrastructure. One can distinguish three levels indicating the extent to which RES can contribute to the total energy supply. These levels are also indicative of the time and changes required to make a significant RES contribution and are defined as follows:

1   maximum level of integration without changing current energy infrastructure
2   maximum level of integration with optimized energy infrastructure to accommodate RES
3   maximum level of integration with a new energy infrastructure.

These three levels should be interpreted rather as a range of shares than as crisp numerical figures. In newly built energy systems, the energy infrastructure can be designed from the very beginning so as to effectively accommodate RES: thus level 3 can practically be reached in the short term. On the contrary, in existing energy systems, with a highly developed energy infrastructure, it will take decades before renewables can play a significant role in the energy supply system. About 40% of the energy is consumed in urban areas with a highly developed infrastructure. For achieving a considerable contribution of RES to the energy balance, serious planning steps have to be taken.

7.4.1.3   Electricity production with RES

Several technical issues are related to electricity production using high penetration levels from renewables.

THE INTERMITTENT CHARACTER OF SOME RE
Considering their technical characteristics, electricity production units can operate either continuously or intermittently. Wind, solar and wave applications produce a rather intermittent energy output, whereas biomass, hydro and conventional units can operate continuously. Obviously intermittent sources cannot reliably cover peak loads, unless the produced energy is stored in a storage system.

It must be noted that biomass plants for electric conversion operate like a conventional unit and so their integration does not affect the stability of the electrical system. On the contrary, for the introduction of wind farms and solar thermal of high capacity in an electrical system it is a prerequisite that suitable studies for both the steady state and the dynamic behaviour of the system be performed.

PROBLEMS ASSOCIATED WITH THE POWER DEMAND SECURITY AND DYNAMIC PERFORMANCE OF HYBRID SYSTEMS
Typical hybrid systems consist of a diesel generator, a photovoltaic generator and possibly other generators such as wind converters and hydroelectric turbines that

complement each other in supplying power. A battery bank and possibly other units for short-term energy storage ensure that power is available at all times. Power is distributed to the loads with AC voltage of the usual frequency and amplitude. Deployment of hybrid systems using exclusively renewable energy sources requires further storage capacity.

In practice the power supply will often include fossil fuel energy sources for support and backup, in particular when hybrid systems are introduced in supply systems with existing fossil fuel power plants.

In order to cope with the large variety of applications and differing power requirements, increasing the reliability and adaptability of the technology is fundamental for promoting the widespread application of hybrid systems using large amounts of renewable energies. Therefore, the broad deployment of hybrid systems using large amounts of renewable energies presupposes improved systems reliability and adaptability to cover various applications and differing power requirements.

ENERGY AND LOAD MANAGEMENT

The operation of power systems with increased renewable energy penetration can be achieved by applying sophisticated algorithms capable of forecasting load and renewable power. The aim should be to maintain a high degree of reliability and security against dynamic disturbances. The development of an adaptable advanced control system will achieve optimal utilization of multi-renewable energy sources, by advising operators of possible actions. Technical constraints imposed by renewable energy sources' availability and variability, as well as thermal power units' technical characteristics, would be reduced by the deployment of an advanced control system, which would ensure the stability of the electrical system.

Rather than attempting to match the power generation to the consumer demand, the philosophy of 'load management' involves an inversion of approach and taking action to vary the load to make it match the power available. When assessing the possible use of load control, consumers' attitudes should be taken into consideration. Consumers should tolerate a complex tariff structure and co-operate in staggering their loads. The incentive is always extremely cheap electricity for the consumers during the hours of low demand, and the result may be a smoothing of the daily power curve, low energy production cost, limited storage requirements and use of a very high proportion of the generated energy.

ENERGY STORAGE OPTIONS

The nature of the electrical load, variability of renewable sources and the characteristics of the electrical grid introduce the need for energy storage. Energy storage devices show the same variety as the technologies for direct use of energy. Energy storage is distinguished as short, medium and long term. From a technical point of view storage technologies can be grouped with respect to the forms of energy being used, mechanical, thermal and electrical. The main available options include batteries, hydraulic/pneumatic accumulators, flywheels and pumped storage. Special applications such as water desalination could also be used as indirect storage.

7.4.2 Non-technical issues

Renewable energy sources have to overcome important barriers, related to the overall deficiencies of the energy market and its mechanisms for technology integration. It is argued that the main non-technical issues influencing the market penetration of RES are information flow, credibility, economics, and administrative and institutional barriers.

In addition to the above barriers, RES integration in the emerging decade has to be accomplished in a new economic environment. The issuing liberalization of the electricity market is expected to further discriminate RES in the market place. Although the experiences recorded in the countries that are ahead in the liberalization process are not enough to draw definite conclusions, it seems that particular attention should be paid to ensure RES deployment in a free but seriously distorted market.

7.4.2.1 *Information flow through society*
In the era of the information society significant gaps exist regarding information about RES. These gaps can be detected in all stages and aspects included in the process of RES implementation, such as data for RES availability in different regions, for technical solutions and for positive and negative side effects. In addition, there is often a lack of information about technology providers, funding sources and administrative procedures.

The lack of information is the consequence of an overall underestimation of RES by all categories of market players: policy makers, public and local authorities, utilities, private investors and the general public. This perception can be inverted only if both state and society are brought into action to improve the current status of knowledge and to disseminate relevant information.

The means to achieve such an information flow are:

- the substantial integration of courses on RES in all educational levels
- the promotion of seminars and other forms of continuous education for engineers and other disciplines
- the systematic collection, recording and elaboration of data about RES potential and of their impacts on the natural environment and on society
- the creation of the necessary supporting mechanisms (central decision support systems, regional energy offices, body of energy engineers, etc.) for providing information and technical assistance to investors and the general public.

7.4.2.2 *Credibility and market acceptability*
Consumers' and investors' reservations – either justified or due to ignorance – are a common obstacle for most innovative technologies. In the case of RES these restraints

are more pronounced since there does not exist a favourable technological and economic environment ready to adopt these innovations. Maturity and credibility are more difficult to attain in a market that is seriously distorted from both a technical and economic point of view.

Problems of reduced credibility can be removed primarily through standardization and qualification procedures. The setting up of standards began in 1995, and great efforts have to be undertaken at an EU and national level. The great variety of RES, their changing characteristics with location and micro-climatic parameters, and the diversity of exploitation technologies, relevant equipment and appliances, are some of the factors explaining delays or inefficiencies of such procedures.

Another reason for the reduced credibility of RE technologies is the malfunction of relevant systems due to the lack of specialized engineers and the general scarcity of reliable information and technical support. The development of educational infrastructures and supporting mechanisms is expected to beneficially influence the perception of energy producers and consumers about RES.

7.4.2.3 *Economics of RES*

Economic issues refer to all the parameters influencing the competitiveness of RES with respect to conventional fuels or energy forms, and are widely considered as the primary factor determining the rates of RES market penetration. There are several aspects to be considered in relation to these issues, which are all interdependent and should be treated through a systematic integral approach.

COSTS AND PRICES

Costs and prices are the main driving forces of the market defining the competition between different energy alternatives. Despite the impressive cost reductions accompanying the technical upgrading and the fast deployment of RE technologies in the market, most large-scale RE applications are still more expensive than competing sources of heat and electricity generation. The expected further drop in production costs along with their market penetration is not likely to make RES able to compete with conventional fuels/technologies in the short to medium term. Besides, such a procedure is very slow, because of the multiple barriers hindering market penetration, among which is the unfair pricing system.

Prices are the signals reflecting the costs of producing goods and their utility for society. If all costs are unaccounted for and utility factors are ignored, then prices give the wrong signal to the market. This is especially true for RES, since their use is associated with significant environmental and social benefits that are not taken into account in market decisions. Moreover, conventional fuels and technologies still benefit – directly and indirectly – from important public subsidies, which further deepen existing cost differences. The widely accepted inefficiencies of the current costing and pricing mechanisms are therefore the main obstacle to overcome in order to establish fair competition rules.

TAKING ACCOUNT OF ENVIRONMENTAL EXTERNALITIES

The use of fossil fuels for heat and electricity production results in substantial environmental deterioration. In particular, atmospheric emissions – which are insufficiently taken into consideration in energy decisions – are widely known to produce serious impacts on the environment. This is because air quality and other environmental assets are considered as public goods and are not priced by the existing market mechanism. Thus, the price of conventional energy sources does not include the costs imposed on society due to the various environmental impacts on human health and on the natural and social environment (e.g. crops, forests, water resources, natural ecosystems, buildings, cultural monuments).

During the 1990s, intensive efforts were undertaken for estimating the external costs associated with energy production and use. All relevant studies (among which is the ExternE project of the European Commission) have proved that environmental externalities of many conventional technologies may exceed the corresponding private costs. The largest components of these external costs are the global warming effect and mortality and morbidity effects due to atmospheric pollutants.

It is clear that RES, which are exempt from all types of atmospheric emissions, are unequally treated in the current market place and that internalization of external costs would significantly alter present perceptions about the relative economic attractiveness of competing energy technologies.

TAKING ACCOUNT OF SOCIAL SIDE-EFFECTS

In addition to their environmental benefits, RES induce other important side-effects, which may affect social welfare in several ways. All these effects depend on a multiplicity of site-specific factors mostly related to the actual configuration of the energy system and the existing infrastructures. Nevertheless, the experience already acquired makes it possible to outline some general trends.

Due to the disparate geographical distribution and low energy flux of RES, their large-scale deployment is expected to lead to the decentralization of the energy system's structure. This shift from the large to the small scale, and from the national to the regional level, induces a beneficial impact on income distribution and regional development. Although the development of RES may in the long term result in a restriction of activities related to conventional energy technologies, existing experiences reveal that relevant losses are minimal in the short to medium term.

Effects on local employment reflect one of the most important aspects of regional development. New jobs will be created mainly during the manufacturing, installation and maintenance of RES projects. This is because RES exploitation is capital intensive, while relevant equipment has a shorter lifetime compared with conventional systems. However, the number of jobs created during the operation phase is relatively low, since the transformation of RES into useful energy occurs with limited human intervention. The creation of new jobs is generally accompanied by the generation of additional income in the region, which directly affects social welfare. In addition new skills and know-how are gradually developing in the region and also produce additional positive side-effects.

The generation of local, regional or national income is a major source of economic inflow to the public finance due to income taxes and social charges. In addition, effects on employment induce a reduction of unemployment compensations. However, the development of RES induces additional economic benefits due to the avoided costs of imported conventional fuels and value added tax on investment cost, maintenance cost and all spin-off activities generated from the implementation and operation of RES projects.

GETTING COSTS AND PRICES RIGHT

Getting the prices right is clearly stated as a precondition for sustainable development in the EU's official documents, including the Amsterdam Treaty. Despite considerable achievements (e.g. the 'ExternE' project) brought about in the assessment of environmental externalities, no progress has been made in the exploitation of the obtained external cost values in energy costing and pricing. Decisions are principally based solely on private costs, and RES continue to make a modest contribution to the EU's energy balance. Incorporation of environmental externalities in existing costing regimes and pricing mechanisms should therefore attain a high priority among the EU's policy goals.

There are several ways of removing market distortions and establishing conditions for fair competition of RES. The most common means for internalizing external costs is the promotion of economic policy instruments aimed at directly correcting pricing inefficiencies. Subsidies have already proved to be an efficient means of compensating RES for the avoided environmental and social costs associated with conventional systems. At the same time, subsidies are capable of motivating market players and indirectly assisting in the overall strengthening of the RES market. On the other hand, taxes and environmental charges are intended to discourage the consumption of polluting fuels and thus to promote clean technologies in the market place. Nevertheless, there is still strong opposition to the wide use of such instruments.

However, it is possible to take account of environmental and social side effects by influencing the costing approaches followed in energy decisions. Changes in the costing regime should at first be envisaged for the electricity sector. As indicated in relevant research studies, incorporating externalities either in planning capacity expansion or in the dispatch procedure will considerably increase the shares of RES without significantly raising the private cost of electricity generation.

FUNDING MECHANISMS

Another factor that greatly affects the economics and overall market conditions for RES development is the origin and provision mechanism of the necessary funds. Funding in the case of RES is not only intended to encourage investors to engage in projects characterized by high capital intensities; it should also aim to reduce the risk associated with RES investments. To this end, new funding mechanisms, such as third party financing, have been shown to be particularly effective.

7.4.2.4 Administrative and institutional barriers

Issues related to the accommodation of RES have long been underestimated. However, it is recognized that even under the most favourable conditions implementation of RES projects may be retarded or hindered due to malfunctions created by administrative procedures and legal loopholes.

The process of issuing installation and operation licences often requires the involvement of several different public authorities and is therefore extremely complicated and time consuming. This can result in considerable delays, which discourage new investors and may even lead to the suspension of the investors' interest. Within the whole licensing procedure, possibly the most serious problems are acquiring the right to use the necessary land and establishing a right of priority to the grid. Simplification and clarification of the whole process is vital.

Other problems may arise concerning the distribution of expenses necessary for grid connection among investors and utilities, as well as the provisions defining the ownership of relevant installations. In addition, legislation should clarify issues related to the right of priority to the grid. This right is of particular importance in the case of autonomous systems and may cause serious complications and delays during RES implementation.

7.4.2.5 Market liberalization

Even before the entry into force of Directive 96/92, setting out the targets and procedures for an internal electricity market in the EU, the electricity industry in many industrialized countries was in the process of a fundamental restructuring. The main characteristic of this process is the rejection of the traditional model of a vertically integrated industry subject to cost-based regulation and the introduction of a free competition regime.

This reform was typically expected to increase productivity, decrease costs and bring about significant benefits to electricity consumers. Besides these positive effects there is to date an increasing concern about the impacts of market liberalization on the development of RES. The reason behind such apprehension is that without removing existing market distortions, it will be even harder for RES to compete with conventional technologies in a deregulated market. Although existing experiences refer to only a few liberalization schemes and relatively short time spans, the outlook is rather discouraging.

In order to prevent such a negative evolution – taking into account the risk of not meeting the obligation of the Kyoto Protocol and the objective of doubling the share of RES to about 12% by 2010 – the Commission has recently elaborated a directive intended to facilitate the penetration of RES in a liberalizing market.

The key issue in the directive is that as long as electricity prices do not reflect the full environmental and social costs of energy sources, the support for RES should be an essential part of the new framework conditions of the internal electricity market. The directive proposes that member states may continue with their national supporting mechanisms until enough evidence is available for establishing a harmonized Community-wide support scheme. Another important issue addressed in this directive is the need to remove all existing administrative barriers hindering the wide implementation of RES projects, as well as the setting out of certification procedures to guarantee the origin of electricity from RES.

This new framework may open new prospects for RES, not only by ensuring the viability of RES projects in the medium term but also by clearly adopting the need to create a level playing field within the internal electricity market, where RES are expected to have a major long-term role.

7.5 GOALS FOR A EUROPEAN RE INTEGRATION STRATEGY

7.5.1 The vision

For Europe the vision in terms of long-term objectives should be:

- to ensure a sustainable energy supply in the long run, i.e. an energy supply concept based almost entirely on RE technologies
- to ensure a controlled and well-planned transition from the present fossil-fuel-based energy supply to a future sustainable one
- to establish a framework for continued, increased RD&D of large-scale application and integration of RE technologies.

The goals, means and strategy necessary to implement the vision are defined in the following sections, and the whole conceptual set-up is shown in Figure 7.1.

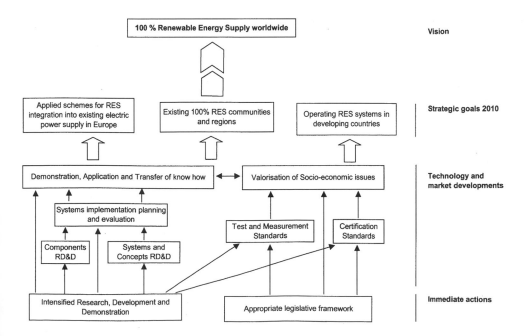

Figure 7.1 Conceptual set-up of the proposed strategy.

7.5.2 Strategic goals

- Development of schemes for integration of RE into existing power supply systems.
- Development of communities and regions in the EU with 100% RE power supply.

These should include development of RE technologies giving the same power supply reliability and power quality as a conventional system.

These regions and systems should be fully operational by 2010. They should be of different sizes and have different characteristics. A certain number should be identified immediately as pilot regions and specific action and implementation plans should be performed.

At the small scale, this could be done using building blocks, new neighbourhoods in residential areas, small rural areas with particular electric and heating needs (e.g. pumping, small manufacturing, water production, refrigeration, etc.), and isolated areas like islands or mountain communities. At a larger scale, several 'solar cities' could be identified where the use of RE in the urban environment could be introduced, larger rural areas with an approach using energy crops as base of a renewable energy activity. Administrative areas could be defined as pilot regions as well as large islands of the Mediterranean (e.g. Sardinia, Crete or Mallorca).

These goals may be indicative, but they should also be clearly stated. As implementation proceeds and experience is gained, the numbers may have to be adjusted: 100% RE-based power supply will be possible in only a few cases, though the time limit of 2010 may turn out to be unrealistic.

In order to reach these goals a number of interdependent targets must be reached, and in order to do this a number of measures must be implemented. To define the actions necessary and to monitor progress a strategy must be defined that includes schedules, priorities and players.

- Development of sustainable and competitive power for developing countries.
  Technology should be developed *and implemented* to meet the needs of the developing world in order to avoid conflicts and the consequences of an increasing use of fossil fuels. This development will ensure that an increasing proportion of third world energy demand will be supplied by RE technologies and will include the development of a competitive technical and economical performance. In the end it will contribute to limiting (the growth in) global CO_2 emissions.

7.5.3 RD&D areas and tasks

RD&D areas and the tasks involved in pursuing the goals are shown in Table 7.1. The measures to be implemented in order to achieve the goals are briefly described. These measures include actions that may also be dealt with as part of the development of specific RE technologies. Some degree of co-ordination between the various technologies should be exercised, so that the various efforts may contribute to each other.

Table 7.1 RD&D areas and tasks

RD&D areas	Important RD&D tasks
Systems and concept studies: • System design and modelling • Component interfacing and interaction • Control and regulation	• Layouts and principles for modular control • Local vs global control • Scheduling and dispatch of system components • Standardized protocols for control communication and power exchange • Tools for flexible modelling and design layout of system configuration and control
Component studies: • Power electronics • Energy storage • Energy carriers	• Converters • Battery technology • Flywheels • Hydrogen storage • Hydrogen in prime movers
Systems planning and evaluation	• Integrated resource planning techniques for integrated RE technology • Quantification of externalities • Capacity values/short-term predictions
Certification and evaluation	• Standards for certification • Standards for evaluation and prediction of technical/economical performance
Test and measurement procedures	• Standards for test, measurements
Applications	• Technology development • Demonstration and technology transfer • Development and deployment in second and third world • countries
Socio-economic issues	• Internalization of energy externalities • Articulation between policies for RES and environmental or sectoral policies • Contribution of RES to innovation and technical change

7.5.3.1 *Develop and demonstrate proven and cost-effective concepts, principles and solutions for RE systems*

In order to reach the goals for application of RE systems both inside and outside the EU, cost-effective concepts and solutions should be developed and demonstrated. In the process there will be a substantial contribution to the international competitiveness of the European industry, and support skills.

Development efforts should include work on configuration and architecture of both concentrated RE systems and dispersed RE schemes, and the philosophies and principles for integration and dispersion would form part of this. Control and regulation, local as well as global, are important issues, which are linked to the interfaces and interactions between components. Solutions applied for specific cases should be demonstrated and monitored.

Preference should be given to activities involving combinations of technology and application in such a way that such projects have the potential to cover the entire range from prefeasibility study through feasibility study and demonstration phase (mainly programme financed) to large-scale implementation with (mainly) commercial international financing.

7.5.3.2 Improve support technologies, e.g. energy conversion and management technologies

In order to improve the ability of RE systems to supply guaranteed and reliable power, support technologies such as energy conversion, storage and management technologies should be developed and improved.

Electrical power conversion technologies and storage technologies, such as battery and flywheel storage, to support the RE generating technologies are known as **hard** technologies, but **soft** technologies, such as management of input (RE resources, e.g. biomass, etc.) and output (consumer loads), should also be included.

An important issue is the development, incorporation and management of secondary loads such as water (including water desalination), heating, cooling, freezing, etc. The generation and storage (and distribution) of energy carriers such as biofuels and hydrogen should be part of this effort.

Ultimately the development and demonstration of the application of energy carriers in prime movers such as power generators and vehicle engines will make a major contribution towards a high utilization of RE-based power supply.

7.5.3.3 Flexible, modular and updateable tools for planning, design and evaluation

A prerequisite for the increased integration of RE technologies into European grids and regions is the acceptance of RE technologies by decision makers, financiers, etc. This means that both technical and economical credibility should be increased and substantiated.

An important part of this will be the development of flexible, modular and updateable tools for planning, design and evaluation of such systems and schemes. This implies the need for tools for the flexible modelling of system configurations and control strategies as well as system planning models and integrated resource-planning techniques. Models should be developed for technical development as well as for techno-economical assessment of performance.

Part of this will be the need for the development of agreed criteria for the evaluation of technical and economical performance, so that proposed solutions may be compared with other alternatives in a generally agreed and accepted way.

7.5.3.4 Ensure necessary quality of deliverables (power, reliability, predictability)

As part of increasing the technical and economical credibility of RE systems and schemes the quality of the deliverables from such systems and schemes should be ensured.

This requires work within areas such as:

- interface and interaction between systems, components and users
- verification and documentation
- monitoring of actual installations.

Examples of deliverables in this context are power, reliability and predictability.

7.5.3.5 Harmonized European standards, legal structures and institutional framework

Another prerequisite for a successful large-scale and high penetration deployment of RES will be to ensure a free operation of the markets for such systems and schemes. A major part of this will be to work towards harmonization within Europe. Examples are:

- standards to improve quality, performance and compatibility, such as standards for:
  - protocols and interface between components and users
  - testing and measurement procedures
  - technical and economical performance
- legal structures such as certification, planning permits, noise limits, etc.
- institutional issues such as insurance, financing, etc.

This will also contribute towards an increased industrial collaboration and co-production with respect to these systems and schemes.

7.5.3.6 Agreed rates and principles for the quantitative evaluation of RE technologies

This is a prerequisite for any widespread acceptance of RE systems and schemes by implementing agencies, donor organizations and other international financiers and decision makers.

The work includes agreements on:

- energy accounting principles and rates
- economic assessment principles and methods
- quantification in economic terms of emission reduction effects and socio-economic effects.

Such agreements will enhance the application of RE systems and schemes both inside the EU and in the developing countries.

7.5.3.7 Proven and competitive applications of integration of RES by 2010

This is very much a matter of implementing conscious and well-planned, executed and monitored integrated applications of RES in the form of actual systems and schemes, i.e. a matter of successful demonstration and application.

In order to obtain the goals in this context it may be necessary to include such issues as:

- subsidies and policies
- establishing manufacturing capabilities
- formation of networks for resources including manufacturers, R&D institutions, consultants, etc.

Some of these issues include components of policy and politics.

7.5.3.8 *Improve knowledge of socio-economic issues and design effective policies*

Understanding the multiple implications of RES development for the overall socio-economic structure and exploiting knowledge for the design of effective policies may significantly enhance their market penetration.

The significance of 'getting prices right' is strongly emphasized in several official EU documents. The important research efforts of the 1990s in the assessment of energy externalities have already provided valuable information, which remains largely unexploited. Although there is a need for continuing research in this area in order to capture additional (less tangible) effects – both environmental (e.g. biodiversity) and non-environmental (e.g. security of energy supply) – and for reducing relevant uncertainties, what is really missing are ways for tracing the incorporation of these figures in the decision- and policy-making procedures. It is therefore necessary to be able to estimate market responses to such changes in order to smooth possible disturbances and to maximize welfare in the short and long term.

In this context, the design of policies for RES should take into account the overall structural changes taking place in Europe. Understanding these changes and their interactions with human behaviour, social priorities and lifestyles could significantly assist in the development of strategies for better accommodating RES in a changing society.

Synergies and tensions between different areas of policy should also be investigated. Interconnection of RES with environmental policies has already been established. This is not the case for policies for regional development and employment, although the existing interactions have been adequately ascertained. Particular attention should be paid to the analysis of the contribution of RES to process innovation and technical change and their impact on the overall competitiveness of the EU industry.

7.5.4 Road to the market: establish the appropriate legislative framework

The analysis highlighted in the previous sections has already highlighted the major factors affecting RES development and the consequences of their market penetration. Despite the differences due to site-specific factors the following general remarks can be derived:

- The current pricing mechanism does not fully reflect all cost parameters associated with the production and use of conventional fuels.
- Most RES usually find it difficult to compete with conventional fuels and energy forms, due principally to higher investment costs.
- RES are environmentally benign, and their use allows major environmental burdens associated with conventional fuels, which negatively affect human society and natural ecosystems, to be avoided.
- RES may enhance regional development, and their net balance regarding the generation of new jobs and social income seems to be positive.

- Society is not well informed about the advantages and disadvantages of RES, while the existing market and administration structures impose significant barriers hindering their massive implementation.

All the above remarks illustrate the major legal gaps concerning the promotion of RES in the market place. A legislative framework capable of removing the above listed distortions should ensure the following:

- Establish corrective pricing measures aiming to 'get the prices right'. This means imposing taxes on conventional fuels and/or subsidizing RES in order to take account of hidden or avoided costs (environmental or social externalities) associated with different energy alternatives.
- Establish efficient funding mechanisms in order to encourage private or public bodies to undertake the risk of investing in RES technologies.
- Set up transparent and fair conditions under which innovative technologies like RES will be able to participate in a liberalized energy market.
- Remove existing administrative and bureaucratic barriers hindering RES implementation in practice, such as licensing of RES projects, conditions for access to the grid, etc.
- Promote the dissemination of technical and non-technical information regarding the use of RES. Special emphasis should be given to introducing relevant courses and seminars in all educational levels.
- Increase the credibility of RES technologies by specifying technical standards and promoting efficient accreditation and control mechanisms.

8. RENEWABLE ENERGY TECHNOLOGIES FOR DEVELOPING COUNTRIES

B McNelis (IT Power), G van Roekel (ECN), K Preiser (FhG-ISE)

8.1 INTRODUCTION

In Europe and industrialized countries in general, there is an established energy production and distribution infrastructure. Taking electricity as an example, there are national and supranational grids with a variety of conventional (including large hydro) generating plants feeding electricity into the grid; while consumers, from individual households to large industry, draw electricity from the grid. Renewable energy technologies are being developed and applied alongside 'conventional', i.e. fossil fuel and nuclear, sources of energy, for electricity, heat and transportation. The renewable electricity generation technologies described in earlier chapters, from small roof-top photovoltaic (PV) installations to multi-MW wind farms, also feed electricity into the grid, thus displacing conventional fuels.

As has been mentioned, for Europe to increase its use of renewable energy, the key issues relate to making these technologies cost competitive and integrating them into the energy system. But, for more than 2 billion people living in rural areas and villages in the world's developing countries, the situation is rather different. These people have no access to electricity, and the distribution chain for liquid fuels is very limited. In some countries, particularly sub-Saharan Africa, women and children spend many hours each day collecting wood or dung to use as cooking fuel, but production, collection and use are very inefficient. Kerosene and candles are used extensively for lighting. The real cost of these is extremely high, but there are no alternatives.

There are important potential synergies between renewable energy resources and the needs of the rural population of developing countries. Moreover, in many developing countries, biomass is already the predominant primary energy resource for cooking, although it is often not used in a sustainable manner. Given these pre-existing conditions, it should be possible for renewable energy technologies to impact on the energy scene without first developing extensive fossil-fuel-based economies and infrastructure.

This short chapter can do no more than highlight the key issues. Hydro, biomass and wind have the potential to make a big impact on energy supply, but their major drawback is that their successful utilization is very site specific. One particular technology, photovoltaics, is considered for two reasons. First, small, simple PV systems are well proven and can be deployed almost universally in the rural areas of the developing world, where their performance is not nearly so dependent on specific site conditions, as is the case for hydro and wind. At present costs, PV is economically competitive with the traditional alternatives: kerosene lamps, candles, primary batteries, small gasoline generators, and extending the grid for small demand levels. Secondly, in Europe there are major R&D programmes, widespread manufacturing and government support programmes to create markets and stimulate investment by industry. The EC's 'Campaign for Take-Off' includes an export initiative for 500,000 PV village systems for decentralized electrification in developing countries. An average installed capacity of 1 kW$_p$ and a total investment in the region of €1.5 billion (for which it has been suggested that one-third will come from public funds) is anticipated.[1]

There is considerable scope, in the near term, for technology co-operation. And in the longer term, if the projected cost reductions are achieved, PV could become a significant energy option in countries with severe shortages, such as China and India.

Over the past 25 years, there have been numerous projects to introduce or demonstrate renewable energy technologies in developing countries, funded by well-meaning donor agencies. There have been some successes but many failures, which have in some quarters given renewable energy a bad reputation. There have been, and still are, quality problems, but failures are due mainly to the small and limited scale of projects, and the inherent problems of the aid process. It is now appropriate, with the accepted knowledge of success and failure, to embark on new initiatives to promote the use of renewable energy technologies for poverty alleviation and sustainable economic development.

8.2 PRESENT SITUATION

This section relates the energy requirements of developing countries to the technologies described in the previous chapters.

8.2.1 Energy demand

Figure 8.1 compares energy consumption for selected regions of the developing world with Europe and other industrialized countries.[2] It shows that per capita energy use in the United States is 80 times higher than in Africa and 40 times higher than in South Asia. If Africa and South Asia were to increase their energy consumption to US, or even European levels, using conventional energy sources, the impact on the global environment would be tremendous. However, these statistics do not tell the full story, as they refer to 'commercial energy'.

More than 2 billion people in the developing countries obtain most of their energy from traditional fuels (i.e. 'non-commercial' fuels, which are not included in Figure 8.1's statistics): wood, crop residues, charcoal and animal dung. These fuels are used

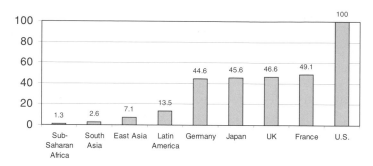

Figure 8. 1   Per capita consumption of commercial energy in selected regions and countries, 1992. (Source: World Bank)

mostly for cooking. In sub-Saharan Africa, for example, as much as 80% of all energy consumed is in this form and for this use. Wood and dung are inefficient energy sources for cooking (gas is 10 times more efficient). These fuels have to be collected, usually by women and children spending many hours per day – time that could be devoted to productive work (e.g. farming) or education, thus addressing the overriding problem: poverty. Moreover, the local environment is damaged when trees are removed for firewood, and breathing the smoke from wood and dung combustion in confined spaces results in respiratory disease and premature death for millions of people.

The developing countries therefore face two enormous problems: how to meet the increasing demand for commercial energy for those with access (principally in the cities), and how to provide access to modern, efficient and clean forms of energy for the majority of the population in rural areas. Renewable energy technologies provide the best solution in many situations.

There have been a variety of studies on and projections of future energy supply and demand, such as those of the World Energy Council[3] and the International Energy Agency.[4] The major oil company, Shell, has made its own analysis, and published a number of likely scenarios.[5] Shell suggests that most growth in demand will, in time, be met using renewables, with renewables providing more than 50% of world energy demand by 2050. After 2050 supplies from fossil fuel and nuclear sources are expected to decline. This scenario is illustrated in Figure 8.2.

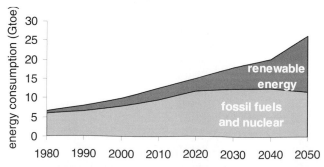

Figure 8.2 World annual energy consumption, Shell Energy 'Sustained Growth' scenario.

It should also be noted that research, development and implementation of renewable energy technologies in Europe and other industrialized countries are driven by concerns about the global environment, particularly CO_2 emissions from fossil fuel combustion. At the Earth Summit in Kyoto in 1997 targets were agreed for reductions in emissions of the six main greenhouse gases, including CO_2, by the industrialized countries (OECD).[6] These ambitious targets will only be achieved by measures that include investment in renewable energy technologies. The CO_2 emissions of developing countries will soon exceed those of developed countries, as illustrated in Figure 8.3. Therefore, development of renewable energy technologies in and for developing countries is required, not only to meet the needs of those without energy, but also to preserve the global environment.

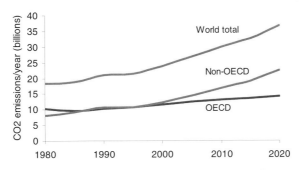

Figure 8.3 Global annual carbon dioxide emissions, according to the IEA, World Energy Outlook 1998.

8.2.2 Renewable energy technologies

It is beyond the scope of this chapter to discuss all the possibilities for renewable energy. Table 8.1 lists the renewable energy options that might be used for electricity, cooking, heating, cooling, water desalination and transport.

Table 8.1 Renewable energy options relevant to developing countries

Solar	PV	Uniquely attractive for individual household electricity Universally applicable Industrial backing
	Water heating	Not priority
	Thermodynamic electricity	Grid-feeding possible, but not yet implemented
Wind	Electricity	Grid-feeding Promising for village grids
	Water pumping	Limited success
Hydro	Electricity Mechanical power	Where resource available least cost for village electrification
Biomass	Cooking	High priority
	Electricity	Grid or mini-grid feeding
	Transport	Biofuels, vegetable oils

8.2.2.1 Biomass

Biomass is already used extensively (32% of developing countries' primary energy, compared with 3% in Europe), and is a major source of employment and income in rural areas. In rural areas firewood is often used for cooking, while in urban areas charcoal is the main cooking fuel. Charcoal production and sales is a multi-billion Euro business and employs millions of people.

In contrast, agricultural residues are a huge resource, which is generally not used for energy production. In many countries (e.g South East Asia, Ghana, Gabon, Cameroon, Côte d'Ivoire), the wood producing industry is an important contributor to local economies. The use of biomass wood residues to generate electricity and thermal power and to support weak grids is therefore important for the economic sustainability of this industry, as well as a key component of sustainable energy demand growth. While promising feasibility work for cogeneration is currently being undertaken in Ghana, with ample replication potential for other countries, there are key implementation barriers still to be overcome. The key ingredients still missing are tailored financial engineering and support for spreading the financial risks for such projects.

The development of new types of biomass stoves has been the subject of research, development and demonstration (RD&D) throughout the world, with work in Europe, co-operating with many developing countries, funded by the donor agencies. In China, 120 million stoves have been supplied to rural areas. In India a national programme has supplied 8 million 'chulhas' with a government subsidy of at least 50%. However, there are indications that half of these are no longer in use. There has been more success

Figure 8.5 Collection, delivery and sale of firewood is big business in urban areas (Sri Lanka). (Photo: I T Power)

Figure 8.4 Collecting firewood is a daily chore for millions of women and children (Indonesia). (Photo: I T Power)

in countries such as Kenya and Ethiopia, where commercial enterprises (as opposed to donors, NGOs or government departments) have sold stoves in response to popular demand.

Charcoal is often the preferred biofuel. It burns without smoke or dangerous flames and it is easier to control. But charcoal must be bought, not collected. There are some indications that in countries or regions where the economy is expanding, families prefer to work for income and purchase charcoal, rather than spend hours of their time searching for and collecting wood. Charcoal production is a major rural industry in many countries, creating employment, but production is inefficient, with much more firewood being used than is necessary. One alternative under investigation is pelletized agroresidue biofuel produced by small, rural, cottage industries.

In China 5.5 million households are reported to be using biogas systems for cooking. These use animal manure, kitchen wastes and 'night soil' as feedstock.

8.2.2.2 Photovoltaics (PV)

PV can make an important contribution to help meet basic needs and increase productive use activities. It is uniquely suited to provide small amounts of electricity (for lighting in particular) for household systems where there are weak grids, for drinking water supply pumping, and increasingly for water irrigation. PV systems are technically proven, and are competitive at present prices. Peri-urban and hybrid PV systems may provide critical core business components alongside rural market segments to local PV entrepreneurs, providing them with much needed diversity in their business development. Families who are able will purchase solar home systems (SHS) with cash, in preference to continually buying kerosene. If suitable financing (e.g. microfinance) can be provided there are strong indications that sales could increase dramatically. There are some millions of small PV systems in daily use, but market penetration is still minimal. There are currently projects in 20 countries supported by €200 million of grants from the Global Environment Facility (GEF), with total costs more than €1 billion, aimed at opening markets for SHS. Significant results are not yet available from these projects.

Figure 8.6 shows a solar home system on a house in Lesotho. This has been purchased with cash by the householders, who do not anticipate a grid connection and are aware that investment in PV provides better quality lighting and is lower in cost than kerosene.

Applications of PV that can have a huge impact on quality of life include water pumping and vaccine refrigeration. PV water pumps provide water on demand from bore holes – a huge improvement on open wells. In the countries of the Sahel, PV pumps have transformed the quality of life and improved health in villages. The European Union financed a €60 million regional solar programme with the Comité permanent Inter-Etats de Lutte contre la Sécheresse dans le Sahel (CILSS), which installed 1270 PV systems in the mid 1990s in Burkina Faso, Chad, Gambia, Niger, Mali, Mauritania and Senegal. This was a large PV programme, but it reached only about 1% of the 67,000 villages in the CILSS countries.[7] The success of this programme has now led to a second phase, with a further €60 million from Europe, which is currently

Figure 8.6 Solar home system, Lesotho. (Photo: IT Power)

Figure 8.7 Solar lighting system for a women's training centre, Sudan. The light means that artisan training can be provided in the evenings. (Photo: IT Power)

in preparation. The European Union has recently provided €17 million for PV to electrify 1000 rural schools in South Africa.[8] Projects to introduce PV for household electrification in South Africa are underway, including an ambitious initiative by Shell. However, a serious barrier is the erroneous perception that solar electricity supply targeted at the rural (black) population is 'inferior' to the grid electricity used by the (white) city dwellers. The rural schools project should go some way to addressing this problem.

New applications for PV are also being further developed, such as PV refrigeration (both mobile and stationary units) for small-scale entrepreneurs in rural areas and for urban use. These developments are responding directly to market demand in areas where PV solar home systems have been successful. However, RD&D support remains a stumbling block to pre-commercial development.

8.2.2.3 Small hydro

The global installed capacity of small hydro is less than 40 GW – half of which is in China. This is believed to be less than 10% of the worldwide economic potential (400–500 GW), the majority of which is in developing countries. Numerous developing countries, particularly in Africa and Asia, have barely touched their small hydro resource.[9]

Figure 8.8 illustrates a 12 kW system in Nepal, which operates off 20 m head and is used by day for rice-hulling, flour-grinding, and ice production. At night it distributes electricity to local households, who pay a fixed monthly connection charge. The scheme is viable only because of the productive daytime activities – electricity generation barely breaks even because of the investment in the transmission lines and the difficulty in collecting the small monthly charges.

Figure 8.8 Micro-hydro installation, Nepal.
(Photo: I T Power)

8.2.2.4 Solar thermal power

Solar thermal power has more potential in developing countries than in Europe. As shown in Chapter 5, demonstration projects are now being developed in Egypt, India, Mexico and Morocco, with support from the GEF. The government of the Indian state of Rajasthan is developing a 140 MW_e integrated solar combined cycle (ISCC) power project. The waste heat from naphtha/gas-fuelled gas turbines will be integrated with the solar thermal energy from a field of parabolic troughs (about 220,000 m^2) to generate steam for a turbine/generator.

There have been attempts to develop small-scale solar thermal power plants. In the 1970s to early 1980s several research companies installed more than 100 systems in several developing countries, using flat plate or parabolic collectors and low-temperature Rankine cycle engines, in the range 1–25 kW. These were not ready for deployment, and development has been abandoned. There are dish Stirling engines (25 kW) currently under development.

8.2.2.5 *Wind*

Wind is already being used in countries with a sharply increasing demand for electricity. As noted in Chapter 6, India has the fourth largest installed wind power capacity in the world (1 GW). Although there were problems with the early installations, the annual growth rate is now around 20%. China, with more than 250 MW installed, is expected to be one of the most significant markets.[10]

Wind technology has developed and has become more cost-effective by increasing the size of turbines. In the mid 1970s a typical wind turbine was rated 30 kW with a diameter of 10 m. Technology development and market pull have led to machines today being rated around 1 MW, with 2.5 MW, 80 m diameter being introduced. Given that the market is for large, grid-feeding wind generators, the European and world industry has focused on meeting this demand. Consequently, small, stand-alone wind generators for village electrification have been neglected. There is very limited availability of 10–50 kW wind generators, but there is clearly the commercial potential for these to be developed and used for village electrification.

Very small wind generators (i.e. micro-wind generators, rated at 50–100 W) have had commercial success (though this is negligible in terms of business volume or installed capacity compared with wind farms), because of the demand from the yachting and leisure markets in Europe and elsewhere. It is claimed that there are 120,000 small-scale wind generators for household use in Inner Mongolia, China. There has been manufacture in (Outer) Mongolia through a joint venture with a European manufacturer.

Figure 8.9 50 W wind generator in use by nomadic family in Mongolia. (Photo: I T Power)

8.2.3 Rural electrification

Over the past 25 years an additional 1.3 billion people in developing countries have been connected to the electric grid – a great achievement; however, world population has increased by 2 billion over the same period. So there are 700 million more people without electricity today than 25 years ago.[11] There are now more than 2 billion people living in 400 million households without access to the electricity grid. The grid will never reach the majority of them. Most developing countries are in crisis dealing with increasing demand from already-connected consumers, and grid extension to low-demand consumers is not bankable. But there are clear indications that the 'wealthiest' of the unelectrified 2 billion are able and willing to pay for electricity if the service can be provided.

Figure 8.10 shows electricity cost as a function of load for PV, wind, small diesel and micro-hydro power generation.

It is very difficult to generalize about renewable energy options, because the resource is site specific. For very small loads, i.e. individual households, wind can be considered the least-cost option, but this is dependent on there being a good wind regime. Where there is a good wind there have been successes with micro-wind generators. PV, however, is much more widely suitable, as solar radiation is adequate in most areas of most unelectrified, populated regions, and the installation requirements are less demanding. In addition, PV modules are mass produced by a number of companies and so there is far greater market penetration for PV than for wind. PV is thus uniquely attractive for individual household electrification.

For larger village-level loads, the most commonly used electricity generator is a small diesel generator supplying a village distribution system. Where a suitable hydro resource is available, however, micro-hydro is less expensive to install and operate. There are many unelectrified villages in developing countries, as well as villages that are using diesel generators, where a suitable river or stream is available for micro-

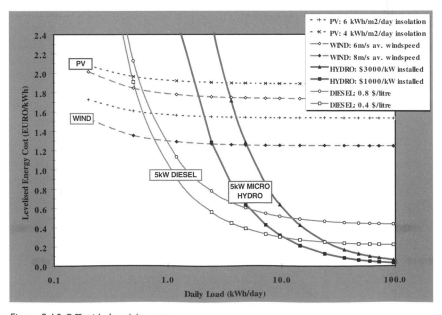

Figure 8.10 Off-grid electricity costs.

hydro to be effectively utilized.[12] This is largely an untapped market outside China, though in China there are already 20 GW in more than 50,000 installations.[9] Micro-hydro can be installed for around €1000/kW, and on the basis of reasonable performance and economic assumptions, can provide a connection for household lighting, TV, etc. at a cost of around €1.7/month. This is easily affordable by many, and such schemes should be readily financeable.

8.3 GOALS FOR RD&D

8.3.1 Biomass

As noted above, wood, agricultural residues and dung are the principal cooking fuels of poor people. In urban areas wood is used but there is a preference for charcoal. As incomes rise, households move to modern fuels, principally kerosene, and LPG in a few countries. Biofuels will remain the principal source of energy for cooking for many years to come, so it is believed that RD&D should focus on improving the efficiency of the use of biofuels, and promoting more sustainable ways to produce and deliver such fuels. Focus should be placed on the conjunction between technology, design and implementation barriers so that results can be achieved in the field. Such work is best performed in the countries affected; however, this should not preclude the provision of expertise from Europe, which is focused on working in synergy with local partners in the field.

8.3.2 PV

Most activity to date with PV installations in developing countries has been R&D, testing and demonstration. Some projects have done more harm than good, and have not addressed the real issues of building sustainable commercial markets.[13] There are already proven and sustainable markets for PV in locations where PV is more convenient and cost-effective than alternative power sources, e.g. remote telecommunications. The same conditions of convenience and cost-effectiveness apply generally to the rural areas of developing countries, but market penetration is minimal. The high cost of PV is not, by itself, *the* barrier to market expansion; rather it is lack of finance, specific development of sustained and locally based micro-credit financing beyond project level, and an enduring infrastructure to bring together the consumers, products and finance and payment collection systems. With more attention to overall implementation research and continued focus on developing the scale of successful projects, PV will move forward even faster. With support at the level given, for example, to PV in Germany, PV could become a significant energy source.

The International Energy Agency PV programme's co-operation with developing countries project[14] is helping develop 'Best Practice' guides with the aim of assisting in opening sustainable PV markets. But a more important aim of the project is to have a 'political' impact. The IEA member countries are also the major international donors, and the project is using the IEA infrastructure and status to try to influence donors to increase their funding allocations for renewable energy. Key requirements are to demonstrate that access to electricity can contribute to poverty alleviation, and that PV is the 'best' means to supply the electricity.

There are still serious problems with the quality of some PV components and systems. This should not delay implementation of new, well-designed and managed PV projects, but urgent attention is required. The Global Approval Programme for Photovoltaics (PV-GAP) is tackling these issues.[15] There is considerable scope for the European PV research community to contribute further to this area. Village PV systems still need to be successfully demonstrated (although the expression 'demonstration projects' should be avoided, and demonstrations as executed in the past should not be repeated).

8.3.3 Small hydro

The most pressing need in small hydro is not for original R&D, but to develop in-country, or at least regional, technical capability in order to bring costs down. Otherwise new projects can proceed only with foreign consultants and equipment, and a huge development loan. The starting point should be at the micro-hydro level (<300 kW), where the risks and complexities are reduced, and the environmental impact is minimal.

8.3.4 Solar thermal power

No specific R&D is foreseen, as large-scale plant RD&D is already well advanced and includes the relevant developing countries.

8.3.5 Wind

As noted earlier, there should be more attention given to the 'technology gap' preventing readily available, commercial wind generators from being used in village electrification. This could be with stand-alone (i.e. with battery) or diesel hybrid configuration. RD&D on this subject is in progress in Europe, but needs to be expanded.

8.4 ROAD TO MARKET

Development of the market for renewable energy technologies in developing countries will be achieved only through co-operation between developed and developing countries on RD&D and technology transfer. This requires the involvement of the European bilateral and multi-lateral donors. But there is a huge educational and promotional effort required to make such institutions aware of, and ultimately supportive of, renewable energy technologies. RD&D and institutional and donor support efforts should include significant facilitation to train and support the development of the private sector initiatives in ways that support but do not create dependence. An initiative on *renewable energy for sustainable development* should be developed and presented to the European authorities. The European Union is a major donor, funding €10 billion annually. Moreover, the EU and its member states provide more than 50% of total world aid. McNelis in *Electricity for All*[16] calls for 10% of this aid to be applied to achieve the goal of *Power for the World,* proposed by Palz in 1994.[17] The renewable energy community (including the EUREC Agency and the European Renewable Energy Export Council) should work together to take such an initiative forward.

REFERENCES

1 European Commission, *Energy for the Future: Renewable Sources of Energy: White Paper for a Community Strategy and Action Plan*. COM(97)599 final, 26/11 November 1997.
2 World Bank, *Rural Energy and Development: Improving Energy Supplies for 2 Billion People*. World Bank, Washington DC, July 1996.
3 World Energy Council, *Energy for Tomorrow's World*. World Energy Council, London, 1993.
4 IEA, *World Energy Outlook*. IEA, Paris, 1998.
5 Shell International, *The Evolution of the World's Energy System*. Shell International, London, 1996.
6 UNFCC, Full text of the Kyoto Protocol, adopted at the third session of the conference of the Parties to the UNFCC in Kyoto, Japan, December 1997.
7 CILSS/PRS, *Regional Solar Programme: Lessons and Perspectives*. European Commission, 1999.
8 J R Bates, J P Louineau, C Pucell, W Mandhlazi, and H Van Rensburg, 'Programme for the electrification of 1000 schools in Eastern Province and Northern Cape in the Republic of South Africa', *Proceedings of the Sixteenth European Photovoltaic Solar Energy Conference*, Glasgow, May 2000, James & James (Science Publishers) Ltd, London, 2000, 3074–3077.
9 O Paish, P Cowley and T Jiandong, 'Small hydro in China: status and opportunities', *Water Power and Dam Construction*, December 1998.
10 EWEA, *Windforce 10*, 1999.
11 J Bond, J, *Opening Statement, World Bank, World Bank Energy Week: Extending the Frontiers of the World Bank's Energy Business*. World Bank, Washington DC, October 1998.
12 O F Paish *et al.*, *Micro-hydro Power: A Guide for Development Workers*. IT Publications, 1991.
13 B McNelis, 'PV rural electrification: needs, opportunities and perspectives', *Proceedings 2nd World PV Conference*, Vienna, July 1998.
14 J R Bates and B McNelis, 'IEA PVPS Task IX: Deployment of photovoltaic technologies: co-operation with developing countries', *Proceedings of the Sixteenth European Photovoltaic Solar Energy Conference*, Glasgow, May 2000, James & James (Science Publishers) Ltd, London, 2000, 2989–2992.
15 PV-GAP, Geneva: c/o IEC Central Office, 3 rue de Varembé, Box 131, CH-1211 Geneva 20, Switzerland. Website: www.pvgap.org
16 B McNelis, 'Electricity for all: the PV solution', *Proceedings of the Sixteenth European Photovoltaic Solar Energy Conference*, Glasgow, May 2000, James & James (Science Publishers) Ltd, London, 2000, 2092–2095.
17 W Palz, 'Power for the world: a global photovoltaic action plan', *Proceedings of the Twelfth European Photovoltaic Solar Energy Conference*, Amsterdam, 1994, HS Stephens & Associates, Bedford, 1994, 2086–2088.

9. OCEAN ENERGY

P Fraenkel (Marine Current Turbines Ltd)

9.1 INTRODUCTION

The oceans cover more than two-thirds of the surface of our planet and represent an energy resource that is theoretically far larger than the entire human race could possibly use. Land-based renewable energy technologies are already facing constraints, for example conflicts over land use, visual intrusion, etc. The seas will therefore be important in the future as they offer huge open spaces where future new energy technologies could be deployed on a grand scale, without serious impact on either the environment or on other human activities.

Unless we develop and use marine renewable energy resources, we are unlikely to be able to meet our energy needs without continuing to burn increasing quantities of fossil fuels; land-based renewable energy resources are unlikely to be able to meet more than a small proportion of future energy needs. This is the main argument for investing in these new and so far little developed energy solutions. However, marine renewable energy resources are generally more costly and difficult to access than the land-based options (which is why experience with them so far has been quite limited).

The huge size of the marine energy resource is to some extent academic as most of the energy available is either too diffuse for economic exploitation with known technologies, or located too far from the markets where it could usefully be applied. However, there are places where the different types of marine energy tend to be concentrated. Fortunately such locations are often close to the shore and in such cases may coincide with a market close enough for any energy that can be captured. In such cases, there are good prospects for future exploitation.

9.2 THE POTENTIAL OF MARINE ENERGY: STRATEGIC SUMMARIES

There are seven quite different marine energy resources that could conceivably be developed, namely:

- offshore wind
- tidal/marine currents
- wave energy
- OTEC (ocean thermal energy conversion)
- tidal barrages
- salinity gradient/osmotic energy
- marine biomass fuels.

Offshore wind is dealt with under the topic of wind energy, as it is derived from the atmosphere rather than from the sea, and so will be discussed no further here. However, there are generic technical problems for offshore wind that also apply to some of the other technologies (e.g. installation techniques, marine cables, piling, many of the legal and institutional requirements, etc.). Therefore, various areas of R&D relating to offshore wind may also be particularly relevant to tidal current and wave energy, and vice versa.

Of the other resources, the only two with serious short- to medium-term prospects for widespread use to meet European energy needs are tidal/marine currents and wave energy; however, for completeness the rest will be summarized as they may offer interesting prospects in the longer term and might therefore be valid topics for research.

9.2.1 Tidal/marine currents

Marine currents are driven mainly by the rise and fall of the tides under the influence of the gravitational fields of the sun and moon, but also by differences in sea water density caused by salinity and temperature variations and by oceanic circulation driven by Coriolis forces caused by the rotation of the earth.

The mechanisms for exploiting this kinetic energy resource are similar in principal to a wind turbine, for example a vertical or horizontal turbine mounted in the flow, driving a generator (see Figure 9.1). An advantage of the marine current resource is that it is generally predictable with a high level of confidence, since the drivers tend to be gravitational rather than weather.

In most places the movement of sea water is too slow and the energy availability is too diffuse to permit practical energy exploitation; however, there are locations where the water velocity is speeded up by a reduction in the cross-section of the flow area, such as straits between islands and the mainland, around the ends of headlands, in estuaries, etc. As with wind energy, a cube law relates instantaneous power to fluid velocity, so a marine current of 2.5 m/s (5 knots) represents a power flux of approximately 8 kW/m². This is far more intense than solar or wind resources, which are commonly used. The minimum velocity for practical purposes is about 1 m/s (2

Figure 9.1 Computer generated image of an array of axial flow tidal current turbines of a kind under development by Marine Current Turbines Ltd in the UK, showing how a system might be maintained by raising it above the sea surface. (© Marine Current Turbines Ltd)

knots), which is around 0.5 kW/m². Hence the main siting requirement is a location with flows exceeding about 1.5 m/s for a reasonable period with sufficient depth of water to cover a reasonable size of turbine (perhaps 15–30 m).[1,2]

Data on marine currents are limited to certain sea areas, so few predictions on the scale of this resource have been made. A recent study[3] funded by the European Commission, which evaluated the tidal current energy potential for 106 locations around Europe, estimated an exploitable resource from those sites of 48 TWh/yr. Various other figures are:

- 70 TWh/yr available from the Sibulu Passage in the Philippines[4]
- approximately 14 GW (equivalent to about 37 TWh/yr) for Chinese coastal waters[5]
- up to 320 MW in the UK by 2010.[6]

There is, however, the potential around the UK to install many gigawatts of tidal turbines. Worldwide the potential is uncertain but obviously has to be much larger.

9.2.2 Wave energy

The energy in waves is also a kinetic energy resource in the sense that waves, which are generated by the interaction of wind with the surface of the sea, cause a circular motion of water in a vertical plain near the surface. This energy can be accessed in a variety of different ways, as explained later in this chapter. However, waves, being generated by wind, tend to be unpredictable in time, although long-term averages at specific locations tend to be consistent.

There are numerous mechanisms for exploiting the energy available in waves,[7] several of which are currently being constructed.

The oceanic wave climate (i.e. far offshore) offers enormous levels of energy; power levels vary from well over 60 kW per metre of wave front in the north Atlantic to closer to 20 kW/m in the less energetic tropical regions. As waves approach the shore, energy is dissipated, leading to lower wave power levels on the shoreline. Therefore, the energy availability is sensitive to location and the distance from the shoreline: e.g. the practicable UK wave energy resource could be 0.4 TWh/yr at the shoreline, 2.1 TWh/yr near shore and as much as 50 TWh/yr offshore.[7] Eventually wave energy could make a major contribution by yielding as much as 120 TWh/yr for Europe[8] and perhaps three times that level worldwide.

9.2.3 OTEC (ocean thermal energy conversion)

OTEC is based on using thermodynamic systems, such as a vapour (Rankine) cycle engine, to extract power from temperature differences between the surface and the seabed in deep oceans. The exploitation of natural temperature differences in the sea by using some form of heat engine has been considered and discussed for the best part of 100 years, and this resource is potentially the largest source of renewable energy of all.[9] However, the laws of thermodynamics demand as large a temperature difference as possible if a technically feasible and reasonably economic system is to be delivered, and this limits the application of this technology to a few tropical regions with very deep water. So there is virtually zero potential for conventional offshore OTEC in European waters.

However, an alternative is to heat sea water on the shore (e.g. in a solar pond) and use the sea as the cool heatsink for the thermodynamic process; this is sometimes known as **shoreline OTEC** and could conceivably be applied on the sunnier European coastlines having relatively cold sea water to act as a heatsink.

An attraction of OTEC is that the sea represents stored solar energy, so any such system can operate night and day and is not subject to significant fluctuations in energy availability. The latest thinking is that OTEC needs to be applied as a multi-purpose technology: e.g. the nutrient-rich cold water drawn from the deep ocean has been found to be valuable for fish farming, also the cold water can be used directly for cooling applications in the tropics such as air-conditioning.[10]

9.2.4 Tidal barrages

Tidal energy has been exploited on a small scale for centuries in the form of tidemills; these are waterwheels similar to those used on rivers, which are placed on a small barrage or dam across a river estuary or at the opening of a lagoon so that they are turned by the tidal flows.

The modern form of this technology depends on creating a large artificial lagoon behind a barrage in a coastal location with a high tidal range; the lagoon can be filled at high tide and the water can be released through turbines as the tide falls. It is in effect a low-head hydroelectric system, but because it can only function intermittently the potential capacity (or load) factor tends to be rather poor (around 20–30%). The barrage usually accommodates bulb turbines, which are necessarily exceptionally large for their rated power due to the high flow required to compensate for the low head. In some cases these can be reversed and used as pumps to give an element of pumped storage.

The tidal barrage has been demonstrated on a large scale (as explained later in this chapter) and investigated in considerable detail, but it is now thought to be economically unattractive, and also many of the proposed schemes would have a major environmental impact. Therefore it seems unlikely that tidal barrages will be developed in Europe in the near to medium term at least.

9.2.5 Salinity gradient/osmotic energy

Salt gradient energy is derived from the latent heat of dilution (water vapour takes more energy to evaporate it from the sea than from a fresh water lake and this inherent energy difference between fresh water and sea water is released whenever freshwater re-enters the seas). The energy released where every river reaches the sea is enormous, at 2.65 MW / m³ per second of fresh water mixing with seawater. This is the equivalent of a waterfall 270 m high at each river mouth. If this energy could be harnessed, its theoretical potential would be enormous; however, it is a very diffuse energy resource and practical technology to exploit it does not exist at present. Therefore it is unlikely that this resource will be applied on any scale for the foreseeable future, although it could be an interesting topic for academic research into possible long-term energy solutions.

9.2.6 Marine biomass fuels

Just as biomass fuels can be cultivated on land, so they can theoretically be cultivated and harvested from the sea in the form of seaweed or kelp; however, it seems unlikely that this will be cost-effective for large-scale energy production for many years.

9.3 PRESENT SITUATION

9.3.1 Tidal/marine currents

Tidal and marine current energy (not to be confused with tidal barrage technology mentioned earlier) is the most recent of the marine energy resources to be seriously studied; most work in this field dates from the last 10 years or so. This is perhaps surprising as the use of water currents for energy is not a new idea (floating waterwheels were used on rivers for this purpose for several hundred years). However, large-scale energy generation from currents requires a totally submerged turbine and to do it reliably offshore requires large and robust systems that are only just becoming technically feasible.

The various turbine rotor options generally coincide with those used for wind turbines because the basic physical principles are analogous. The two main types are the horizontal axis, axial-flow turbine (i.e. with a 'propeller' type of rotor) (as shown in Figure 9.1) and the cross-flow or 'Darrieus' turbine in which blades rotate about an axis perpendicular to the flow. In general the more promising rotor configuration seems to be the conventional axial-flow rotor (as with wind energy conversion).

The maximum flow velocity tends to be near the sea's surface, so marine current turbine rotors ideally need to intercept as much of the depth of flow as possible, but especially the near-surface flow. Options for securing a rotor include: mounting it beneath a floating pontoon or buoy; suspending it from a tension leg arrangement between an anchor on the seabed and a flotation unit on the surface; and seabed mounting (feasible in shallow water, but more difficult in deeper water).

There have been a number of short-term demonstration schemes:

- a floating 3.5 m diameter axial-flow turbine, which delivered 15 kW in Loch Linnhe (Scotland) in 1994[2]
- a 3 kW turbine on the seabed off the Japanese coast for 9 months in 1988
- a floating system of about 5 kW in Australian waters[11]
- a range of vertical axis devices (up to 100 kW) were tested in the 1980s in Canada and the USA.[12]

Work is underway on several demonstration and commercial schemes: a 300 kW grid-connected, horizontal axis tidal current turbine in the UK[13] and a 250 kW vertical axis demonstration scheme in Canada.[2] There have been a number of important European projects in this field, notably:

- *UK Tidal Stream Review for DTI (1992–1993).* Indicated a gross UK tidal current resource of 50 TWh (EPDC, Sir Robert McAlpine, Binney and I T Power).
- *10 kW proof of concept project (1992–1994).* Generated 15 kW with 3.5 m rotor in 2.3 m/s current (Scottish Nuclear, I T Power and NEL).
- *EU-Joule 'CENEX' Project JOU2-CT93-0355 (1994–1995).* Study of exploitation of marine currents in Europe (Tecnomare and I T Power) – 106 potential sites analysed – aggregate capacity of 12,500 MW yielding 48 TWh/yr.
- *Tidemill feasibility – Orkney and Shetland – EC (1994–1996).* Feasibility study for tidal current power in Orkney and Shetland (ICIT and I T Power).
- *'Seaflow' Project – EC Joule – (1998–2002).* World's first grid-connected experimental tidal current turbine – 300 kW – in the UK (I T Power, Seacore, ITT Flygt and ISET-University of Kassel).

• *'Optcurrent' Project – EC Joule – (1998–2001)*. Development of techniques for tidal current energy resource assessment (Robert Gordon University, University College Cork, I T Power).

9.3.2 Wave energy

Most wave energy is to be found in deep water far offshore, and a generic problem with wave energy conversion is that the extreme loadings in severe storms can be very much higher than the 'design conditions', demanding huge reserves of strength. For this reason, virtually all of the few systems that have run for any length of time are shoreline devices: i.e. built into the shore, where conditions tend to be less severe.

The most popular shoreline device is the OWC (oscillating water column), which consists of a large chamber with a submerged opening to the sea – see Figure 9.2. The chamber encloses an air volume, which is compressed by the wave pressure. The air from this space can be driven in and out of the chamber in response to the rise and fall of the waves through an air turbine system, the Wells turbine, which is capable of generating electricity from this reversing air flow (research is in hand with variable-pitch Wells turbines and variable-pitch impulse turbines). OWCs do not necessarily need to be fixed to the shoreline or the seabed; floating OWCs have been demonstrated, notably in Japan. So eventually OWCs can also be considered for deepwater applications.

The alternative to the OWC is a family of devices sometimes known as **point absorbers** (Figure 9.3), which mainly work by having a buoyant component that can react against either an anchor on the seabed or an inertial mass so as to generate power (usually hydraulically). The point absorber (PA) concept can occupy a small space yet capture the energy from a larger area surrounding it using resonance effects. The primary technical problem with most devices of this kind is to provide something for the device to react against (i.e. to absorb the incoming forces from the waves). The point absorber has the advantages of small size (cost and size tend to be related) and suitability for modularized line production.

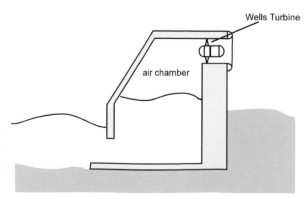

Figure 9.2 Oscillating water column shoreline wave energy converter – air is expelled and drawn in by the wave motion, and this oscillating flow drives the unidirectional Wells turbine.

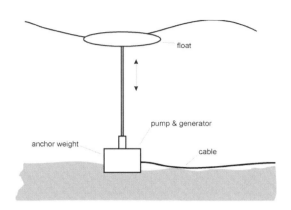

Figure 9.3 A schematic of a typical point absorber wave energy device – there are many variations on this theme, which shows a buoyant vessel connected to a pump and generator mounted on the seabed – movement of the vessel pump sea water generates electricity.

Total grid-connected wave power at the present time is about 1 MW, consisting of several small shoreline-based OWC devices in various countries and several larger OWCs under construction in the Azores,[14] the UK[15] and Australia,[16] together with a floating OWC in Japan.[8] A new generation of point absorber devices is currently under development and due to be installed in the next two years.[8] Taken together, these schemes will increase the world's wave energy capacity by about one order of magnitude during the next few years.

There are several European wave energy projects under development:

- Three commercial schemes in the third round of the Scottish Renewables Order.[14,17,18]
- Demonstration OWC schemes.[13,14]
- Commercial point absorber schemes,[8,19] three under SRO, two under EC Joule, etc. (to be completed similarly to tidal currents).
- Wave energy programmes in Denmark, Portugal and the UK.

There have been numerous EC-sponsored projects:

- JOUR0132, Wave studies and development of resource evaluation methodology (1992–1993)
- JOUR0133, European pilot plant study (1992–1993)
- JOU20394, Offshore wave energy converters (1994–1995)
- JOU20390, Atlas of wave energy resource in Europe (1994–1996)
- JOU20314, European wave energy pilot plant on the island of Pico, Azores, Portugal (1994–1996)
- JOU20283, The deployment and testing of a prototype Osprey wave energy converter – Phase 1 (1994–1998)
- JOU20315, Electricity generation by pilot realization of a wave energy converter (1994–1995)
- JOU20276, A European wave energy pilot plant on Islay (UK) (1994–1996)
- JOR3950002, Making a variable-pitch turbine and high-speed valve for the Azores

oscillating water column (1996–1998)

- JOR3950012, European wave energy pilot plant on the island of Pico, Azores, Portugal. Phase Two: Equipment (1996–1997)
- JOR3950009, The detailed design, manufacture and commissioning of a prototype WOSP wind/wave energy plant (1996–1998)
- JOR3971004, Wave energy device: broadband sea power energy recovery buoy (1997–1998)
- JOR3980312, Islay wave power plant (1998–2001)
- JOR3980282, Performance improvement of OWC power equipment (1999–2000)
- JOR3987026, Wave energy device – broad band sea power energy recovery buoy (1999–2000)
- ERK5-1999-20001, Establishment of a European thematic network on wave energy (wave energy network) (2000–2003).

9.3.3 OTEC (ocean thermal energy conversion)

There are two main processes used for power production from this source, both based on the Rankine (steam/vapour) cycle:

- **The open cycle system**. Evaporated warm sea water is flushed into vapour (at reduced pressure) and then drawn through a turbine by condensing it in a condenser cooled by cold sea water. This produces fresh distilled water when it is condensed, which is potentially a useful by-product.
- **The closed cycle system**. Warm sea water is used to boil a low-temperature fluid such as ammonia, which is then drawn through a turbine by being condensed in a heat exchanger with cold sea water and then recycled back to the boiler by a feed pump.

Offshore OTEC is technically difficult because of the need to pipe large volumes of water from the seabed to a floating system, the huge areas of heat exchanger needed, and the difficulty of transmitting power (and perhaps desalinated water) from a device floating in deep water to the shore.[20]

Although the resource is probably the largest renewable energy source available to us and is therefore practically inexhaustible, it has the major disadvantage of being a peculiarly difficult resource to exploit cost-effectively with known technology due to the large size and poor efficiency of the required plant. It seems unlikely that offshore OTEC will make a major contribution to energy needs in the foreseeable future.

9.3.4 Tidal barrages

The only large, modern example of this technology is the 240 MW$_e$ La Rance scheme, built in France in the 1960s, which is the world's largest tidal barrage by far. An 18 MW tidal barrage system was commissioned at Annapolis Royal in Nova Scotia,

Canada, in 1984 and two systems each of about 0.5 MW have been built in Russia and China. Numerous studies have been completed for potentially promising locations with unusually high tidal ranges, e.g. the Bay of Fundy in Canada, and the €12 billion, 8.6 GW scheme for the Severn Estuary in the UK, but the UK government decided not to proceed with it. Although various smaller schemes have been investigated, the combination of high costs, major environmental impact and poor load factors makes this technology unattractive. Suitable sites for such projects are quite rare, and most schemes of this kind have proved to be extremely costly. So there seems little prospect for much future development of tidal barrages in the short to medium term, except in especially promising locations, so most plans to do so have been abandoned.

9.3.5 Salinity gradient/osmotic energy

This process still has a long way to go even from small-scale demonstration at sea. The following physical principles might be applied:[21]

*   **Osmotic pressure**. A semi-permeable membrane is placed between sea water and fresh water so that an osmotic pressure difference can be generated across it that could be used to generate power. However, this is far from straightforward to engineer at this time.
*   **Electrochemistry**. Here reverse electrodialysis would involve alternate cells of sea water and fresh water separated by electrically charged membranes with properties that would cause an electric current to flow between the membranes due to the passage of salt ions from sea water to fresh water, which give up their charges on the way. This electric current could be used as an energy source. In practice there is no known form of electrochemistry that can be applied economically and practically on the huge scale that is needed.
*   **Differential evaporation**. The difference in vapour pressure between fresh water and sea water can theoretically be used by placing a turbine between two chambers in which vapour is flash evaporated from the fresh water and, in effect, condensed into the chamber containing sea water.

In most cases the efficiency of these processes could be enhanced by interacting fresh water with brine that has been concentrated from sea water, perhaps by using solar energy (e.g. salt-pans).

9.3.6 Marine biomass fuels

Similar technological processes could be used to generate power from marine biomass as from land-based biomass. If a combustion process is to be used, then considerable energy input would be needed to dry the raw material, so perhaps wet processes (e.g. anaerobic digestion) might be most practical. Little work has been done in this field so far.

9.4 ECONOMIC ASPECTS

Because of the limited experience with marine renewables it is difficult to be certain of how economic they might be if developed to a mature stage. There is experience (albeit limited) with tidal barrages, but their failure to 'take off' speaks for itself. In addition, in many cases the economics of these technologies are site dependent (e.g. distance to shore, speed of marine current, average wave power levels, etc.). Reviews in the UK of renewable energy technologies[7,22] generated load factors and costs for some of the offshore technologies, as shown in Table 9.1 (except where other sources are indicated).

It should be noted that several of the options in Table 9.1 could already be commercially competitive in the context of island communities using conventional small-scale diesel generation, which typically can cost in the range from 10 to as much as 50 c/kWh. Hence some of the marine renewable energy technologies may find their initial role in niche markets of this kind.

Table 9.1 Present status of marine renewable energy technologies

Technology	Maturity	Load factor (%)	Installed capital cost (€/kW)	Unit cost of electricity (Eurocent/kWh)
Tidal barrage[22]	Mature	20–25	4000–5000	10–13 cent
Wave – shoreline OWC[8]	Demonstration (Commercial – 2002)	26	2100	~10 cent
Wave – near-shore OWC[8]	Demonstration (Commercial – 2003)	29	1500	~8 cent
Wave – offshore – point absorber[8]	Demonstration (Commercial – 2005)	34–57	1800–3000	4–10 cent
Tidal current turbine[8]	Demonstration (Commercial – 2005)	21–25	1800–2100	4–10 cent[1]
OTEC	Research (Demonstration – 2005)	80 + ?	Not clear	20+ cent
Salt gradient	Not feasible	80 + ?	Not predictable	–
Marine biomass	Not feasible	80 + ?	Not predictable	–

9.5 ENVIRONMENTAL ASPECTS

The environmental impact of nearly all the offshore renewable energy technologies tends to be minimal. Most are unlikely to have much impact except in their immediate vicinity. The exception to this is tidal barrages, where the creation of a large artificial brackish lake behind the barrage changes the estuarine environment radically and has the potential to affect fish and bird breeding and feeding, siltation, etc. None of these technologies produces significant atmospheric or water-borne pollutants. All except marine biomass are carbon negative.

The main issues for all the technologies tend to be conflicts with other users of the seas, e.g. with fishing, marine traffic, leisure activities, etc.

9.6 GOALS FOR RESEARCH, DEVELOPMENT AND DEMONSTRATION

9.6.1 Generic goals for primary offshore technologies

Any programme to develop the technology needs to be supported by a range of activities in three main areas as follows:

*   resource exploration and assessment
*   engineering studies (experimental and theoretical)
*   economic and market studies.

There are common factors for the three main offshore technologies that could make a significant contribution to European energy needs in the foreseeable future, namely offshore wind, tidal/marine currents and wave energy. These are:

*   development of lower-cost techniques for deployment/installation/maintenance of the systems at sea – which needs the development of specialized vessels such as jackup barges
*   development of lower-cost techniques for laying marine cables and for protecting cables from disruption
*   development of design and operational standards requisite to unmanned marine structures (current requirements for manned oil and gas platforms would be too onerous).

The various technologies under development also have specific RD&D needs.

9.6.2 Realistic goals for tidal/marine current energy development

9.6.2.1 Short-term goals (immediate)
Pilot projects to test and demonstrate marine current turbines.
   The main short-term goal should be to test and demonstrate several pilot projects to gain operational experience, to find out the true costs, and to confirm the assumptions behind feasibility studies so far completed.
   There has already been identified a need for two types of system, one that could be seabed mounted and surface piercing for relatively shallow seas (20–30 m depth) such as in the UK, Ireland and Northern France, and one for deeper waters such as in the Mediterranean and other oceanic locations. The latter might be suspended from a buoy or vessel or be seabed mounted but totally submerged.
   The practical turbine size for pilot projects would be from 5 to 10 m diameter with ratings in the 20–100 kW range.

9.6.2.2   Medium-term goal (by 2005)
The development and demonstration of the first small field of grid-connected marine current turbines in the 200–800 kW size range, to test and demonstrate workable technology.

9.6.2.3 Long-term goal (by 2005–2010)

The long-term goal is to see the large-scale commercial take-up of marine current technology (i.e. installation of several fields of turbines aggregating hundreds of megawatts of installed capacity).

Some key technical targets to be achieved are:

*   long periods between maintenance and m.t.b.f. (accessing a marine current turbine is likely to be difficult and costly)
*   corrosion resistance combined with durability
*   effective sealing of enclosures
*   good efficiency over a broad range of velocities
*   reliable interconnection between individual turbines and with the shore
*   secure and economical methods of deployment and recovery
*   development of more advanced turbine designs.

9.6.3 Realistic goals for wave energy development

9.6.3.1 Short-term goal (immediate actions)

*   The construction and operation of a number of first generation (OWC) pilot plants on the European coast over a period of about 3–5 years.
*   Supporting research into optimization of devices and design methods.
*   The technology development and construction of a number of second-generation (point absorber) WECs.
*   Further research on the development of technologies for offshore WECs (e.g. moorings, linkages, bearings, electrical connections).
*   The development of efficient power conversion systems for all devices.

9.6.3.2 Medium-term goal (by 2005)

*   The construction and operation of multi-megawatt demonstration plants based on the first generation devices (OWCs).
*   The construction of pilot power plants based on arrays of second-generation (point absorber) WECs.
*   The development of technology to allow pilot plants for third-generation (larger offshore) devices to be deployed.

9.6.3.3 Long-term goal (by 2010)

*   The resource available for wave power converters in Europe is large and therefore not a barrier to potential development. It would be reasonable to expect wave energy to realistically contribute about 1 TWh per year as this represents only 1% of the technical wave energy resource.

- The installed capacity of WECs is to be 200 MW with an annual installation rate of about 20 MW (largely first- and second-generation devices.)
- The demonstration of third-generation offshore devices.

9.7 ROADS TO A STRONGER MARKET

The route to the market lies through programmes of R&D combined with demonstration projects. There will also be a need for the parallel development of enhanced and less costly techniques for installing and servicing offshore technologies.

At the last analysis ocean energy is available in prodigious quantities; the key question is whether it can be accessed, converted to a useful form and delivered cost-effectively in comparison with other methods of energy generation. In summary the situation is that:

- marine current energy is only just starting to be experimented with, but it involves less technical risk than wave energy (conditions are less extreme) so it could develop relatively quickly
- wave energy is beginning to see success with shoreline systems. Some small-scale, near-shore schemes have been demonstrated, and their effectiveness and economics should be demonstrated over the next 3–4 years. Larger-scale near-shore and offshore schemes have yet to be demonstrated at full size
- OTEC is being experimented with but is of little relevance to Europe except as a possible export product for use in tropical locations. Shoreline OTECs may be more easily applied economically than devices floating in deep waters and may have relevance to southern European coastlines having plenty of sun and cool adjacent waters
- tidal barrages have been tried in a limited way and abandoned without further study, largely because they are perceived at present as uneconomic
- the two remaining options, exploiting salinity gradients and the cultivation of marine biomass, seem to be a long way from practical application and at present are only being considered experimentally.

In the long run the more promising of these technologies, especially wave and current energy, seem certain to be needed if future energy needs are to be based primarily on renewable, non-polluting energy resources. They can be developed at different rates depending on the level of interest and support they receive from governments.

REFERENCES

1 P L Fraenkel, Marine currents: a promising large clean energy resource', *Proceedings Power Generation by Renewables*, Institution of Mechanical Engineers, London, 15–16 May 2000.
2 IT Power and Tecnomare, *Non-Nuclear Energy Joule II Project Results: The Exploitation of Tidal Marine Currents*. Report EUR 16683 EN, DG Science, Research and Development, European Commission, Luxembourg, 1996.

3 *Exploitation of Tidal Marine Currents.* Project JOU2-CT94-0355, EC DGXII, 1996.
4 T Carstens, *A Global Survey of Tidal Stream Energy.* SINTEF, Norway, 1998.
5 Wang Chuankun, *Introduction to Regionalization of the Marine Energy Resources in Chinese Coastal Rural Areas.* 2nd Chinese Oceanographic Institute, Hangzhou, 1989.
6 *New and Renewable Energy: Prospects for the UK for the 21st Century: Supporting Analysis.* ETSU Report R-122, Harwell, UK.
7 T Thorpe, 'An overview of wave energy technologies: status, performance and costs', *Proceedings Seminar on Wave Power: Moving Towards Commercial Viability*, Institution of Mechanical Engineers, London, 30 November 1999.
8 A Lewis, 'Ocean energy', in *The Future for Renewable Energy: Prospects and Directions*, EUREC/ James & James (Science Publishers) Ltd, London, 1996, 58–63.
9 G Boyle (ed.), *Renewable Energy: Power for a Sustainable Future*, Open University and Oxford University Press, 1996.
10 NREL, 1999.
11 http://ee.ntu.edu.au/ntcer/projects/tidalpower/main.html
12 http://www.bluenergy.com/history/
13 'SEAFLOW' PROJECT – EC Joule – (1998–2002): world's first grid-connected experimental tidal current turbine – 300 kW – in the UK (I T Power, Seacore, ITT Flygt and ISET-University of Kassel).
14 A F O Falcao, 'Design and construction of the OWC wave power plant at the Azores', *Proceedings Seminar on Wave Power: Moving Towards Commercial Viability*, Institution of Mechanical Engineers, London, 30 November 1999.
15 Heath and Whittaker, 'The history and status of the Limpet Project', see reference 13.
16 http://www.energetech.com.au/
17 Yemm, 'The history and status of the Pelamis wave energy converter', see reference 13.
18 Lagström, 'Sea power international: floating wave power vessel', see reference 13.
19 Van Breugel, 'The challenge of mobilizing finance for funding the construction of wave power plants in Europe'.
20 SERI, 1989.
21 Charlier, 1993.
22 *An Assessment of Renewable Energy for the UK.* ETSU Report R-82, HMSO, London.

10. SOLAR PROCESS HEAT

K Hennecke, B Hoffschmidt, W Meinecke (DLR),
M Blanco (CIEMAT)

10.1 INTRODUCTION, POTENTIAL AND STRATEGIC SUMMARY

According to a widespread conception, the use of solar energy is associated with the production of either electricity or low-temperature heat, mainly for domestic purposes. This view neglects the great potential for concentrating solar technologies to provide process heat efficiently in a temperature range of 80–250°C and a unit power range between 100 kW_{th} and 10 MW_{th}, as needed in industrial or municipal applications. Typical examples are food, textile and paper industries, auxiliaries for the automobile sector, tanning, curing of building materials, heating and cooling of large building complexes (hotels, shopping centres etc.), and district heating systems. This significant demand cannot be satisfied by any other renewable source of energy, except in locations where biomass or geothermal energy are abundantly available.

Recent studies indicate that lightweight parabolic trough systems can provide heat at costs between €0.03 and €0.1 per kWh, depending on the temperature level and the location in Southern or Central Europe. In the investigated temperature and power range, trough systems can outperform the more commonly known types of solar collector in terms of the annual energy collected and the specific cost of solar heat (Figures 10.1 and 10.2). This may not yet be fully competitive with the still low cost of process heat from fossil-fired facilities and as waste of conventional electricity generation, but it can be attractive for pioneer markets where environmental concerns or a 'green' image justify the incremental costs of the solar system.

In industrial applications, secure energy supply to the production lines is of utmost importance. Therefore, hybrid solar/fossil fuel systems will be preferred initially, starting with small solar shares and full conventional back-up. Batch processes with variable energy demand are particularly suitable to cope with the intermittent solar energy supply. Solar shares can be increased where the energy demand coincides with the daily insolation profile, which is often the case in cooling applications or single shift operations. In cogeneration systems, solar energy can be integrated to reduce heat

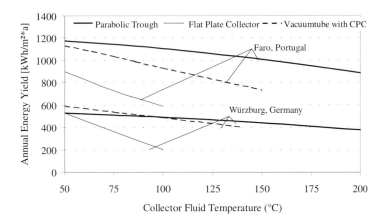

Figure 10.1. The potential annual energy yield of flat plate collector, CPC-vacuum collector and small-sized parabolic trough collector as a function of the mean collector fluid temperature at locations of low and high solar insolation (Würzburg/Germany and Faro/Portugal), as a result of a comparative study, 1998/99 (Source: DLR, Köln)

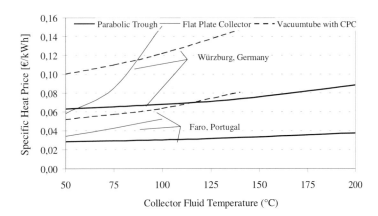

Figure 10.2 The thermal energy generating costs of flat plate collector, CPC-vacuum collector and small-sized parabolic trough collector as a function of the mean collector fluid temperature at locations of low and high solar insolation (Würzburg/Germany and Faro/Portugal), as a result of a comparative study, 1998/99 (Source: DLR, Köln)

demand from the additional boiler, where fossil fuel is used with low exergetic efficiency. In the long term, solar-driven cogeneration systems could become viable, drawing on experience from solar thermal power systems. Similar synergies can be expected from the development of suitable heat storage systems to increase the solar share.

10.2 ACHIEVEMENTS, PRESENT SITUATION, MAIN BARRIERS AND VISION FOR THE WAY FORWARD

It is obvious that a large potential market segment for solar process heat utilisation exists, particularly in sunny locations. But up until now no commercial demonstration of these systems has been achieved.

Until now, the main emphasis has been placed on single-purpose systems using small, low-cost parabolic troughs:

- The most intensive R&D work for parabolic trough collector systems, including field tests, took place during the years 1977–1982. These systems had the capability to supply energy in the low- to medium-temperature range of about 100–300°C, and were designed for applications for both solar electricity generation and industrial process heat production. Key trough manufacturers in this period were US companies like ACUREX and Solar Kinetics (SKI), and the German company MAN. All of these systems were shut down and dismantled after the project operation time.
- From 1983 to 1991, the US companies SKI and IST started new industrial initiatives for solar process heat applications for the market segment of industrial and municipal purposes in the southwest of the USA. These systems use an advanced low-cost, small trough technology with a design power output in the range 100–3,000 kW_{th} for the generation of hot process water up to 150°C. These facilities, with a total installed collector area of more than 9,000 m², are still in operation and have successfully demonstrated commercial operation over more than 10 years.
- In 1998 DLR, mainly in co-operation with IST, started R&D work in order to promote further progress in low-cost, small process heat trough technology, and to build on the USA's progress in process heat troughs. An 80 kW_{th} test facility was installed at the DLR test site in Köln, using an IST low-cost trough collector field (Figure 10.3). This

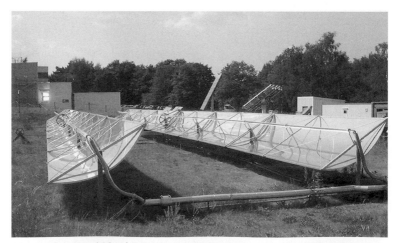

Figure 10.3. The 168 m² IST trough collector system with 80 kWth output at the DLR test site in Köln (Source: DLR, Köln)

system serves as a demonstration for potential applications, and as a benchmark for their own developments of improved components or innovative concepts like the so-called Fix-Focus Collector (Figure 10.4).

- Recently, CIEMAT in Madrid has studied options for thin metallic membrane parabolic trough collectors, resulting in the design and construction of a prototype (Figure 10.5).
- In 1999, the Company SOLEL (Israel) launched its new small parabolic trough collector for process heat production.

Figure 10.4. The 36 m² parabolic trough collector prototype (called Fix-Focus trough) with volumetric air receiver at the DLR test site in Köln (Source: DLR, Köln)

Figure 10.5. Prototype of thin membrane parabolic trough collector designed and manufactured by CIEMAT in Madrid (Source: CIEMAT)

10.2.1 Main barriers

Despite the significant application potential and the promising development status already reached by solar process heat systems based on concentrating collectors, with commercial products already available in the USA and Israel, widespread use of these technologies in Europe is hampered by the following facts:

- Potential improvements of European developments cannot be pursued vigorously, because the required budget is not available at R&D institutions and cannot be afforded by the small companies generally involved. Therefore, no commercial supplier for such systems exists in Europe.
- Potential industrial customers are not sufficiently aware of the development status and possible applications of solar process heat systems, owing to the lack of demonstration projects.
- The high cost of real estate in industrial areas often prevents the installation of large collector fields on the ground. Optimized systems for installation on or integration into industrial roof spaces are not yet available.
- Solar process heat systems need to be tailored to the individual application. Optimized designs have to consider specific site conditions and exploit the potential for rational use of energy in the customer's industrial processes. Numerical tools for efficient preliminary design and assessment of different integration concepts are needed to reduce the engineering effort required in each individual case.
- In many cases, potential customers of solar process heat do not have access to the necessary venture capital. Suitable financing or contracting schemes need to be established to overcome this barrier to market introduction.

10.2.2 Vision for the way forward

A concerted action of demonstration and further R&D will be required to overcome the present vicious circle. Pioneering applications should be realized, if necessary on the basis of imported technology, to demonstrate the potential of the technologies available, to increase awareness by industries and the general public, and to motivate potential manufacturers to produce such systems under licence agreements. In parallel, research and development for competing European commercial products should be supported. Growing competition will spur the market and activate resources for the development of further improvements, which, in turn, will lower the cost of solar process heat and help to open additional market segments.

10.3 GOALS FOR RESEARCH, DEVELOPMENT AND DEMONSTRATION (RD&D)

10.3.1 Successful demonstration of current technology

In the near term, applications with high visibility should be targeted, to improve awareness and motivate additional clients. Preferably, these applications should also show a good coincidence of heat demand and availability of solar radiation.

The tourist sector offers good opportunities in that respect. Hotels in the Mediterranean require most energy for hot water and air-conditioning during the afternoons of the holiday season, when sun is abundantly available. Owing to the large turnover of hotel guests, solar installations will be particularly prominent. Energy costs are only a small fraction of the overall budget in the tourist sector, and the incremental costs of the solar system may be easily offset by the gain in reputation in the eyes of environmentally concerned visitors.

Specific market research should be carried out to identify additional pioneering applications. Relevant market segments should be addressed with specific information describing the state of available technologies and their potential for the targeted application. Several installations should be realized, and their performance monitored for some years. The lessons learned from the evaluation will be important for the development of improved components, materials and system designs.

Both market research and the implementation of demonstration projects will initially require significant public funding support to overcome the barriers presented by lack of awareness and low fossil fuel costs.

10.3.2 Development of numerical design and assessment tools for solar process heat system integration

The large variety of potential applications for solar process heat demands the development of a systematic, computer-assisted approach to:

* analyse the industrial processes with regard to potential for rational use of energy, cogeneration and identification of the most appropriate temperature level and power range for the integration of solar process heat
* assist in the generation of basic design variations for different integration options
* simulate operation of the system to evaluate annual performance, to quantify solar contributions, and to develop and demonstrate suitable operation and control strategies
* evaluate the economic aspects of the investigated design concepts.

In order to efficiently exploit the potential of solar process heat supply, the related industrial processes should be optimized regarding their specific heat consumption. Selection of the appropriate temperature level has a strong influence on the efficiency of solar heat production, and the required power range defines the size of the solar field, which has a major impact on the overall investment. In recent years, methods for process integration (e.g. pinch point analysis) have been developed, but mainly for continuous processes. In the case of solar applications, the variable insolation needs to be considered in optimized designs, and the existing methods need to be upgraded accordingly. Software tools to simulate operation and performance of solar thermal systems are available, but suitable program modules for components used in industrial process heat applications (e.g. specific types of fossil fired boilers, autoclaves, sorption chillers etc.) are still missing. It should be noted that the development and application of these tools will also ease the optimization of conventional, discontinuous processes, opening additional routes to reduce industrial CO_2 emissions.

10.3.3 Development of improved components and subsystems

The main task here is the development and demonstration of collector systems that are easily adapted to different design requirements such as temperature level, heat transfer medium (thermal oil, water, steam, air), or power range. Elevated installation on roofs or above parking spaces etc. will require specific collector designs that minimize wind forces on the collector and the supporting structure. This kind of system should be easily retrofitted to existing buildings. Considerable savings could be achieved with collectors designed to be integrated into roofs or facades of new buildings, replacing and taking on part of the tasks of traditional roof or facade cladding.

Any of these designs will take advantage of newly developed reflector materials, such as anodized aluminium mirrors, thin back-silvered glass mirrors or front surface silvered mirrors. Highly selective absorber coatings and improved receiver designs (e.g. with integrated secondary concentrators) will be developed, enabling increased efficiency at reduced costs. These developments will also create and profit from synergies with similar efforts in the field of solar thermal power plant technologies.

10.4 ROADS TO THE MARKET

In order to reach the necessary synergy of the technology development and market awareness/expansion/acceptance, the road to the market segment of solar thermal heat utilization is projected on the success of current R&D activities for process heat parabolic troughs and on the low-risk approach to advance the state of the technology. Step by step advancement should strengthen the potential customer's confidence in the technology and should avoid frustration due to risky changes in given production processes. Consequently, the use of the well-proven low-cost, small, low-weight collectors is applicable for first demonstration plants. In parallel, improvements of the existing technology and also new approaches (e.g. the Fix-Focus collector of DLR in Köln and the thin metallic membrane collector of CIEMAT in Madrid) are topics for further investigations on the existing test sites, focusing on the following items:

- For fast market introduction the minimum useful installed field size per plant unit might be scaled down to below 100 m². For that reason, the aperture size of the collector unit will probably be decreased.
- The design concepts of process heat facilities using a hot water, hot air or thermal-oil loop should be analysed for comparison, assessed and classified with a view to specific applications. Therefore the replacement of thermal oil by direct steam generation must be analysed and investigated in a process heat trough facility.

- Due to the fast-changing market for process control units, new and more cost-efficient control units have to be selected and adapted for solar use.
- The solar system performance and reliability during representative operating times must be proven for all evident changes.
- Systematic investigations should indicate improvements of the system integration in order to reduce parasitic loads, optimize start-up procedures, and advance the control strategy and the absorber flow control by automatic means.
- Solar heat generation systems using trough collector fields should be optimized for the linkage to conventional single-purpose process heat facilities or to cogeneration plants.

The short-term strategy is to involve interested manufacturers in all further development processes of single components or new approaches in order to be cost-efficient and to achieve an active benchmark.

It is assumed that, within the next decade, solar process heat applications outside some market segments will need to be co-financed by public grants or other sources. Thereafter, due to increasing production rates of collectors, solar systems are expected to become competitive when compared with conventional systems. In the transition period, small systems (with smaller than 100 m² fields) should be available with relatively low absolute capital plant costs, even if the specific costs are higher, to serve a market of ambitious customers who also like to use the system for environmentally oriented advertising.

With the technology already available today, solar heat costs of approximately €0.03/kWh are achievable at favourable locations in southern Europe. In the medium- to long-term, the installed costs of parabolic trough collectors are expected to drop well below €200/m² at high production rates and by implementation of direct steam generation, making them fully competitive when compared with conventional systems.

BIBLIOGRAPHY

Abdel-Dayem, A, Meyer-Pittroff, M, Ruß, W and Mohamad, A., *A Feasibility Study of Solar Heating in Milk Processing*, Eurosun, Portoroz, Slowenia, 1998.

Benz, N and Ruß, W, *Solar Process Heat in Breweries and Dairies*, Eurosun, Portoroz, Slowenia, 1998.

Brown, H, Hewett, R, Walker, A, Gee, R and May, K, 'Performance contracting for parabolic trough solar thermal systems', *Energy Engineering*, Vol. 94, No. 6, 1997, pp. 33–44.

Hennecke, K, Krüger, D and Meinecke, W, 'Integration of solar energy into industrial process heat and co-generation system', in *Proceedings of the 9th IEA-SolarPACES International Symposium on Solar Thermal Concentrating Technologies*, Odeillo, France, 22–26 June 1998 and 1999.

Industrial Solar Technology Systems (IST), *Putting the Sun to Work*, leaflet on IST parabolic trough technology and applications, Golden/Denver, CO, 1997.

Köhne, R, Oertel, K and Zunft, S, 'Investigation of control and simulation of solar process heat plants using a flexible test facility', *Solar Energy*, Vol. 16, No. 2, 1996, pp. 169–182.

Krüger, D, Hennecke, K, Pitz-Paal, R, Hafner, B and Schwarzer, K, 'Performance of a parabolic trough system at low temperatures in different climate', in *Proceedings of the ISES Solar World Congress*, Jerusalem, 4–9 July 1999.

Morales, A and Ajona, J I, 'Durability, performance and scalability of sol-gel front surface mirrors ans selective absorbers', in *Proceedings of 9th SolarPACES International Symposium on Solar Concentrating Technologies, Journal de Physique IV*, Vol. 9, 1999.

Rojas, E, *Membrane Collector Manufacturing and Testing: DISS Phase I Project Final Report*, CIEMAT, 1999.

Rojas, E, 'Metallic parabolic trough collectors: stretched and non-stretched', in *EuroTrough Project First Annual Report*, INABENSA, 1999, Ch. 15.

Schweiger, H, Farina Mendes, J, Benz, N, Hennecke, K, Prieto, G, Cusí, M and Goncalves, H, 'The potential of solar heat in industrial processes: a state of the art review for Spain and Portugal', in *Proceedings of the EUROSUN Congress*, Copenhagen, 19–23 June 2000.

11. SOLAR CHEMISTRY

K-H Funken, C Sattler, C Richter, R Tamme (DLR), J Blanco (CIEMAT), J Lédé (CNRS-LSGC)

11.1 INTRODUCTION, POTENTIAL AND STRATEGIC SUMMARY

Solar chemistry opens up the possibility of integrating renewable energy directly into industrial production. Pioneering experiments in solar chemistry were carried out in the 1970s and early 1980s, particularly in France. The level of research and development in the direct utilization of solar radiant energy increased during the 1990s.

There is a strong interdependence between the energetic use of fossil fuels and their use as raw material for the manufacture of numerous consumer goods. Assuming shortages in oil and natural gas supply in the future, a strong competition between their energy and non-energy use must be expected. Thus it is sensible to decouple the two functions of oil and gas. The most important objective of solar chemistry is to replace the energy function of fossil fuels in the chemical, petrochemical, metallurgical and related industries. If solar radiation is used to produce a fuel identical to or with similar properties as the fuel based on a fossil raw material and a conventional conversion process, the chemical energy sources, such as synthesis gas, hydrogen, methanol, etc., are also called 'solar fuels'.

For the transportation sector the goal is the production of solar fuels, which in the long term could replace oil and natural gas-based products. For photochemical and photocatalytic applications the aim is to substitute expensive artificial light with solar radiation, thus avoiding expenditure on electricity, cooling and light sources. Applications would be possible in a wide field of recycling processes as well. Industrial wastewater can be detoxified using solar photocatalytic technology, and particular problems relating to non-biodegradable contaminants can be solved. The solar sterilization of water without the addition of chemicals like chlorine would also be possible. In future, solar high-temperature process heat could be applied to the recycling of secondary metals, like aluminium, or for the vitrification of filter dusts. Altogether, a wide range of applications for solar chemistry is expected.

11.2 ACHIEVEMENTS, PRESENT SITUATION, MAIN BARRIERS AND VISION FOR THE WAY FORWARD

In the European Union, in particular in France, Germany, and Spain, technological developments ranging from laboratory to small-scale demonstration have been introduced for the qualification of solar chemical processes. There have also been developments in Australia, Israel, Japan, Russia, the United States and Switzerland. These relate to the reforming of natural gas, the splitting of water, the combined high-temperature high-flux destruction of hazardous wastes and recycling of secondary raw materials, the thermochemical conversion of carbonaceous materials, metallurgical processes, the high flux treatment of materials, photochemical syntheses, and the photochemical destruction of contaminants in wastewater and gases. Solar chemical applications use equipment that was originally developed for solar thermal power production, such as heliostats, central receivers, solar furnaces, parabolic troughs, and various concepts of reactors.

Engineering and demonstration scale has already been achieved in particular cases. The reforming of natural gas was the first process for the solar conversion of hydrocarbons, which was tested on an engineering scale of more than 100 kW solar power input.

- At the beginning of the 1990s DLR, Germany, and CIEMAT, Spain, jointly tested solar steam reforming of methane at the Plataforma Solar de Almería (PSA) in southern Spain. The experiment used existing technology, which can be scaled up without major developments. The 170 kW reformer was heated convectively by hot air (1000°C, 9 bar), which was delivered from a 270 kW tubular ceramic receiver.
- DLR, Germany, and Sandia National Laboratories (SNLA), USA, jointly tested solar carbon dioxide reforming of methane in a volumetric receiver-reactor at over 100 kW input on a parabolic dish concentrator in Lampoldshausen, Germany.
- A sodium reflux heat-pipe receiver-reformer was constructed by SNLA, USA, and tested at the 15 kW solar furnace of the Weizmann Institute of Science (WIS), Israel. The feasibility of the approach was confirmed.
- Directly irradiated tubular reformers have been tested at 480 kW scale at WIS, Israel.
- Engineering-scale tests of solar carbon dioxide reforming of methane in volumetric receiver-reactors were continued in German–Israeli (DLR, WIS) co-operative projects SCR (solar central receiver) and SOLASYS (novel solar assisted fuel driven power system). The SOLASYS project is funded by the European Commission within its JOULE III programme and the receiver concept is shown in Figure 11.1.

Within the Solar Energy Association, North Rhine-Westphalia, Germany, Task 3 'Solar Chemistry and Solar Materials Research', several projects have been carried out with a view to developing methods for the solar production of chemicals and high-temperature processes. On a mini-plant scale new receiver-reactors have been constructed and tested mainly in the solar furnace at DLR, Germany. Thus open and closed melt-pot receiver-

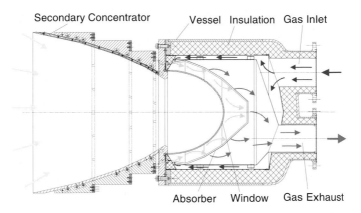

Figure 11.1 Scheme of DLR/WIS high-temperature pressurized volumetric receiver reactor for reforming of methane. (Source: DLR-Stuttgart)

reactors have been developed by the University of Dortmund on a mini-plant scale. The reactors can be used for a wide variety of processes that involve a high-temperature melt phase, such as the thermal production of calcium carbide or the remelting of aluminium scrap. A directly solar heated rotary kiln mini-plant has been developed and tested by DLR. The feasibility of the technological approach is being assessed for the remelting of aluminium, the depletion of volatile non-ferrous metals from residues, and the vitrification of filter dusts. A directly solar heated aerosol receiver-reactor mini-plant has been constructed by DLR for the recycling of waste sulphuric acid.

Photochemical reactions offer access to several industrially important products. Although they yield high value added products with excellent selectivities under mild reaction conditions, their introduction into industrial practice sometimes suffers from the high cost of investment, operation and maintenance. By contrast, solar photochemistry does not need expensive lamps and has significantly lower operation and maintenance costs. Although a solar photochemical plant can only be operated during sunshine hours, the costs of solar technology can compensate for the costs of conventional technology if the reactions require visible light in stoichiometric or sub-stoichiometric amounts. Solar technology for synthetic photochemical applications was mainly established in Germany and Spain.

- The feasibility of solar photochemical technology was demonstrated by German researchers in the SOLARIS (solar photochemical synthesis of fine chemicals) experiment at the PSA in Spain.
- The Max-Planck-Institute of Radiation Chemistry, Germany, tested several photochemical reactions and developed non-concentrating reactors for solar chemical reactions.
- The PROPHIS (parabolic trough collector for organic photochemical syntheses in solar light) facility at DLR, Germany (Figure 11.2), is the most versatile equipment

Figure 11.2 The PROPHIS facility at DLR in Köln involves a two-axis tracking parabolic trough collector with a total reflective area of 30 m² and a maximum solar output of 30 kW. (Source: DLR-Köln)

in the world for the assessment of solar photochemical syntheses. Depending on specific conditions up to 1 ton per year could be converted. The technological approach was extended to many classes of synthetically valuable reactions.

• RWTH Aachen and DLR developed line-focusing holographic concentrator-reactors. Selective solar irradiation yielded a significantly lower demand for cooling energy.

• The SOLFIN (solar synthesis of fine chemicals) test plant at the PSA uses compound parabolic collector technology for solar photochemical applications. Several European groups carried out numerous tests to demonstrate the efficiency of the system for performing preparative synthetic reactions.

• In the DLR solar furnace second-generation solar photoreactors have been tested by DLR and RWTH Aachen with the aim of increasing the space/time yields by allowing operation at high to very high irradiances. This approach was applied both to the production of cyclohexanone oxime, a bulk product for which compact and high intensity operation is required, and to the short wave photochemical syntheses of fine chemicals that use only a small part of the solar spectrum.

Throughout the world, purification of polluted water is a key issue in environmental protection. In particular radiant assisted treatment of polluted waters can be applied successfully to deal with those pollutants that can be degraded only very slowly, or not at all, by conventional procedures, e.g. biological treatment. Many of the sunny regions of the world have an enormous demand for purified water. In solar water treatment the sun and solar equipment substitute for electricity, lamps and related equipment. Research in the field of solar photocatalytic detoxification of wastewater and gases is progressing in many laboratories in Spain, Germany, Portugal, Italy and France. In France, research is conducted mainly by five CNRS laboratories. In some EU-funded projects (AVICENNE and INCO programmes) partner organizations of non-EU Mediterranean states have been involved, including Algeria, Israel, Malta, Morocco, Syria, Tunisia and Turkey. Prototype reactors are being tested, e.g. for the mineralization of pesticides below the standards for drinking water, the treatment of olive oil mill wastewater, or the decolorization of textile wastewater. A water detoxification pilot plant was constructed within the BRITE-EURAM III SOLARDETOX (solar detoxification technology for the treatment of industrial non-biodegradable persistent chlorinated water contaminants) project by partners from Spain, Portugal, Italy and Germany at a company near Madrid (Figure 11.3).

Other fields of interest include:

- the production of metals and energy carriers by the solar thermal reduction of metal oxides; a thermochemical cycle relying on the direct splitting of ZnO is being studied at CNRS-LSGC Nancy, France, and at the Paul Scherrer Institute (PSI), Switzerland
- the solar thermochemical conversion of solid carbonaceous materials – mainly biomass – into upgraded products (gases, bio-oils, char) by gasification or pyrolysis processes (CNRS-LSGC Nancy, France)
- the CNRS-IMP at Odeillo, France, in collaboration with the University of Montpellier, France, is working on the solar production of fullerenes and carbon nanotubes. CNRS-IMP is also working on the solar thermal surface treatment of materials such as the hardening of metals
- the radiant assisted development and testing of ceramic materials, e.g. composite mullite-based ceramics, silicon carbide, silicon nitride, was carried out by DLR and the Universities of Bochum and Duisburg, Germany.

Figure 11.3 The SOLARDETOX pilot plant near Madrid with collector field aperture of 100 m² and plant capacity of 3 m³/d of industrial wastewater throughput, containing chlorinated hydrocarbons. (Source: DLR-Köln)

The further development of solar chemical technologies will have to focus on proving long-term reliability. Another problem that has to be solved involves the optimal interaction between the solar radiation and the reactor in competition with efficient heat transfer and hydrodynamic phenomena. These research efforts relating to the general field of solar chemical reaction engineering will be decisive for economic efficiency.

11.3 GOALS FOR RESEARCH, DEVELOPMENT AND DEMONSTRATION (RD&D)

Generally, the goals are to develop the tools of solar chemical reaction engineering and to establish and prove the technical feasibility and the economic reliability of pioneering industrial applications, which can be considered as reference cases for a broader field of applications in the future. More specifically, the main objectives of solar chemical RD&D are:

- to manufacture solar fuels containing stored solar energy, which could advance to be the 'fuels of the future'
- to detoxify hazardous wastes and contaminants and to recycle materials
- to provide a substitute for electricity in photochemical, photocatalytic and electric arc driven processes being used for the synthesis of commodities such as caprolactam, or of high value added products such as fragrances, fullerenes and carbon nanotubes
- to identify solar-specific reactions and to optimize operation conditions, in particular concerning the interactions between solar radiation and fixed or circulating particles
- to prove the long-term reliability of solar operation
- to detail the design of process-specific solar reactors
- to adapt solar high temperature–high flux reactors to extreme conditions and to select specific materials
- to elaborate an appropriate match of discontinuously operating solar unit processes and conventionally operated unit operations that require steady-state conditions
- to allocate knowledge on the physical properties of materials at high temperature–high flux conditions
- to elaborate the costs of solar chemical processes and compare them with conventional processing.

There are numerous potential applications for concentrated solar radiation in the chemical and related industries. To initiate solar chemical engineering in the context of limited budgets it would be best to concentrate on a number of key processes that could open the door to a broader range of applications. The experience gained in the key processes would give guidance for further developments. In prioritizing approaches in research and development, concentrated sunlight should fulfil more functions than simply replacement of a fossil fuel. Radiant heating with concentrated sunlight is a key issue because there are potential technological advantages when compared with conventional high performance heating processes:

- The heat flux densities that can be achieved with concentrated solar radiation are more than one order of magnitude higher than those available in conventional heating.
- Concentrated radiation impinging on the reactants is located inside a limited volume, the surrounding of which is much colder. This situation appears to be favourable for rapid cooling of the intermediates formed and hence for isolating specific products.
- Radiant heating is free of matter as opposed to conventional methods of applying very high heat fluxes to reactants such as electric arcs or high performance burners. Thus the reactants and the reaction products will not be contaminated by electrode burn-off or the products of combustion. In radiant heated processes the cost of product or exhaust gas purification may be significantly lower than in conventional processes.
- The use of solar photoreactors instead of conventional photoreactors or lasers may be more cost-effective because the installation of the conventional equipment is expensive and it has high O&M costs.

There is only limited knowledge available on the design of solar receiver-reactors, the operation of solar chemical processes and the costs of solar chemical processes on an industrial scale. Proof of feasibility of innovative solar receiver-reactors is required both on a mini-plant and on a demonstration scale. Further R&D should prove exemplary approaches in areas that are relevant for whole branches of the chemical and related industries:

- solar radiant heated high-temperature treatment of hazardous wastes and recycling of secondary raw materials such as aluminium, zinc or mineral acids
- solar radiant heated high-temperature conversion of carbonaceous materials directed towards the production of upgraded fuels and chemicals, including reforming, gasification, pyrolyses and cracking of different feedstocks
- solar radiant heated high-temperature metallurgical processes, e.g. for the production of primary zinc
- high flux material processing and treatment.

11.4 ROADS TO THE MARKET

Most of the anticipated applications need good to excellent climatic conditions with a high availability of direct insolation. But for some applications the availability of a special infrastructure is more important than optimum climatic conditions. In these cases the profit margin of the respective process is controlled by the energy cost only to a limited extent or even not at all. Decisive reasons for using a concentrating solar technology are that either solar equipment is cheaper than, e.g., high-power lasers or

Table 11.1 Potential applications of solar radiation for chemical, metallurgical and related processes

Solar application	Solar technology required for industrial-scale application	Regions of interest	Readiness to market	Cost of solar process compared with conventional processes
Reforming of natural gas	SCR	Sunbelt	Proof of technology in demonstration scale	Appr. double
Thermochemical heat pipe: reforming – methanation	SCR	Coupling of sunbelt regions to industrialized regions	Proof of technology in demonstration scale	Appr. double
Thermochemical heat pipe: ammonia cycle	SCR	Coupling of sunbelt regions to industrialized regions	Laboratory scale, design studies	?
Coal gasification	SCR	Sunbelt	Laboratory scale, design studies	?
Biomass gasification	SCR	Sunbelt	Laboratory scale	?
Oil shale gasification	SCR	Sunbelt	Laboratory scale	?
Biomass pyrolysis	SCR	Sunbelt	Laboratory scale	?
Calcination of limestone	SCR	Sunbelt	Small engineering scale	?
Production of calcium carbide	SCR	Sunbelt	Design studies	?
Reduction of metal oxides for thermochemical water splitting	SCR	Sunbelt	Laboratory scale	?
Reduction of zinc oxide	SCR	Sunbelt	Laboratory scale	?
Generation of fullerenes, carbon nanotubes, and other special materials	SF	Special locations, e.g. Köln, PSA, Odeillo	Laboratory scale	?
High temperature–high flux treatment of metallic and ceramic surfaces	SF	Special locations, e.g. Köln, PSA, Odeillo	Laboratory scale	?
High temperature–high flux testing of metallic and ceramic surfaces	SF	Special locations, e.g. Köln, PSA, Odeillo	Laboratory scale requirements	Depending on specific
Photochemical synthesis of caprolactam	SF or PT	Mediterranean to sunbelt climate	Laboratory scale, design studies	25% cheaper
Photochemical synthesis of fine chemicals	SF, PT, CPC, NCC	Central European, Mediterranean to sunbelt climate	Engineering scale, design studies	25–50% cheaper
Photocatalytic water treatment	CPC, NCC	Mediterranean to sunbelt climate, in particular cases perhaps in central European climate	Demonstration scale, design studies	Same cost level or cheaper for selected applications
Photocatalytic air treatment	NCC	Mediterranean to sunbelt climate	Laboratory scale	?
High temperature–high flux treatment of hazardous wastes	SCR, SF	Mediterranean to sunbelt climate	Laboratory scale	?
Recycling of secondary aluminium	SCR, SF	Mediterranean to sunbelt climate	Miniplant scale in SF	25–100% more than conventional
Recycling of waste mineral acids	SCR, SF	Mediterranean to sunbelt climate	Laboratory scale	?

that solar equipment can be operated in a clean environment as compared with conventional equipment where, e.g., in electric arcs the electrode burn-off contaminates the products. Photochemical and photocatalytic applications appear to be less sensitive to a high availability of direct insolation: thus they may also have a chance in the central European climate.

The majority of potential solar chemical applications have only been investigated at the laboratory scale. For an eventual opening of markets significant efforts have to be undertaken both in proof of feasibility and in assessment of the costs. Small-scale feasibility tests covering low and high concentrating units have to be performed, covering the temperature range from ambient to temperatures up to 2000°C. For selected processes, demonstrations with industrial participation will pave the way for large-scale industrial realization. For the long-term success of solar chemical processes their reliability and their costs are decisive. To encourage implementation of solar technologies for commercial use, further evaluations have to demonstrate the economic advantages and risks.

BIBLIOGRAPHY

Becker, M and Funken, K-H (eds), *Solare Chemie und Solare Materialforschung*. C F Müller Verlag, Heidelberg, 1997.

Blanco, J, Malato, S, Fernández, P, Vidal, A, Morales, A, Trincado, P, Oliveira, J C, Minero, C, Musci, M, Casalle, C, Brunotte, M, Tratzky, S, Dischinger, N, Funken, K-H, Sattler, C, Vincent, M, Collares-Pereira, M, Mendes, J F and Rangel, C M, 'The SOLARDETOX technology', in *Proceedings 10th SolarPACES International Symposium on Solar Thermal Concentrator Technology*, 8–10 March 2000, Sydney, Australia, pp. 201–208

Funken, K-H, Pohlmann, B, Lüpfert, E and Dominik, R, 1999, 'Application of concentrated solar radiation to high temperature detoxification and recycling processes of hazardous wastes', *Solar Energy*, Vol. 65, 1999, pp. 25–31.

Funken, K-H and Ortner, J, 1999, 'Technologies for the solar photochemical and photocatalytic manufacture of specialities and commodities: a review', *Zeitschrift für Physikalische Chemie*, Vol. 213, 1999, pp. 99–105.

Lédé, J, 'Solar thermochemical conversion of biomass', *Solar Energy*, Vol. 65, 1999, pp. 3–13.

Lédé, J, and Ferrer, M, 'Solar thermochemical reactors', *Journal de Physique IV*, France, Vol. 9, No. 3, 1999, pp. 253–258.

Lédé, J and Pharabod, F, 'Chimie solaire dans le monde et en France', *Entropie*, Vol. 204, 1997, pp. 47–55.

Wörner, A, and Tamme, R, 'CO_2 reforming of methane in a solar driven volumetric receiver reactor', *Catalysis Today*, Vol. 46, 1998, pp. 165–175.

APPENDIX A. THE EUREC AGENCY

THE EUROPEAN RENEWABLE ENERGY CENTRES AGENCY

The European Renewable Energy Centres Agency was set up in 1991 as a European Economic Interest Grouping to provide a forum for interdisciplinary co-operation between many of Europe's most respected renewable energy research organizations. It now has over 40 members, ranging from academic institutions and national research centres, to other organizations active in research, education, training, and technology transfer activities. EUREC is an independent association uniting more than 2000 scientists and engineers working in the field of renewable energy. These research fields include:

- biomass
- photovoltaics
- small hydro
- solar buildings
- solar thermal power stations
- wind
- solar process heat and solar chemistry
- ocean energy
- integration of renewable energy
- renewable energies in developing countries

EUREC's activities are as follows:
- EUREC encourages contact and co-operation between its members and European industry as well as EU political groupings keen to promote the use of renewable energy.
- EUREC provides targeted information to the European Commission and the European Parliament.
- EUREC assists members in fast and detailed information transfer from the EC and in the formation of strong groups for European RD&D projects
- EUREC prepares consensus position papers written by its members. These are updated at regular intervals and take into account the views of industry. It also plans roadmaps and strategies for the RD&D of different renewable energy technologies.

- EUREC manages and co-ordinates projects related to the deployment of renewables. This includes the creation, regular updating and distribution of information and preparatory research on international guidelines and standards.
- EUREC is responsible for disseminating know-how of renewable energy technologies to both the general public and specific target groups. In particular, EUREC aims to promote education and training at university and at the medium technical level. It has achieved this through the creation of a European Master's Degree in Renewable Energy Technologies and high-level professional courses which it runs regularly.
- EUREC provides a platform for common initiatives with international organizations and is particularly keen to help transfer knowledge and technologies on renewables to the developing world.

EUREC-Agency EEIG
Renewable Energy House
26, rue du Trône
1000 Brussels, Belgium
Tel: +32 2 546 19 30
Fax : +32 2 546 19 34
Website: http://www.eurec.be

Contact:
Mr. K. Derveaux, Secretary General
Mr. K. Faïz, Office Manager

President (1996–2002)
Vice President (1996-2002)

Prof. J. Luther
Prof. A. Zervos

Former Presidents:

Prof. R. Van Overstraeten †	(1994–1996)
Prof. W. H. Bloss †	(1991–1992)
Prof. G. Peri	(1993)

EUREC Executive Bureau since July 2001:

Prof. A. Luque	Photovoltaics
Prof. O. Lewis	Solar Buildings
Mrs. M. Delgado	Solar Thermal Power Plants
Prof. D. Infield	Wind Energy
Prof. D. Mayer	Integration
Prof. K. Blok	Biomass and Developing Countries
Dr. J. Bates	Small Hydro Power, Ocean Energy and Developing Countries

EUREC Executive Bureau (1996-2001):

Dr. M. Becker	Solar Thermal Power Stations
Mr. J. Beurskens	Wind Energy
Dr. P. Fraenkel	Small Hydro Power and Ocean Energy
Prof. E. Koukios	Biomass
Professor J. Luther	Solar Buildings
Mr. McNelis	Developing Countries
Dr. H.A. Ossenbrink	Photovoltaics
Professor A. Zervos	Integration

EUREC MEMBERS

For the latest information on EUREC and its members, please visit our website at http://www.eurec.be

AEE – Arbeitsgemeinschaft Erneuerbare Energie, Werner Weiss, Post Box 142, Feldgasse 19, 8200 Gleisdorf, Austria

ARMINES – Ecole des Mines de Paris, Didier Mayer, BP 207, 060904 Sophia-Antipolis Cedex, France

CESI – Centro Elettrotecnico Sperimantale Italiano, Gabriele Botta, Via R. Rubattino 54, 20134 Milano, Italy

CNRS Laboratoire Phase, Jean-Claude Muller, BP 20, 23 rue du Loess, 67037 Strasbourg Cedex, France

Conphoebus, Beniamino Morgana, Passo Martino, Zona Industriale, Casella Postale, 95030 Piano d'Arci, Italy

COSTIC – Comité Scientifique et Technique des Industries Climatiques, Eric Michel, Rue Lavoisier, Z.I. Saint Christophe, 04000 Digne-les-Bains, France

CRES – Centre for Renewable Energy Sources, Mrs. Simantoni, 19 km Athinon Marathone Ave., 19009 Pikermi, Greece

CREST – Loughborough University, Centre for Renewable Energy Systems Technology, David Infield, Angela Marmont Renewable Energy Laboratory, Loughborough University, Loughborough, Leicestershire, LE11 3TU, United Kingdom

DER-CIEMAT, Renewable Energy Department, Luisa Delgado Medina, Avenida Complutense 22, Edf. 42, 28040 Madrid, Spain

DEWI – Deutsches Wind, Jens Peter Molly, Ebertstrasse 96, 26382 Wilhelmshaven, Germany

DLR – Deutsches Zentrum für Luft- und Raumfahrt e.V., Solar Energy Technology, Manfred Becker, Linder Höhe, 51147 Köln, Germany

E²H – Energie-Environnement-Habitat, Nadine Levratto, Alain Louche, c/o University of Corsica, Centre Scientifique et Technique de Vignola, Route des Sanguinaires, 20000 Ajaccio, France

ECN – Netherlands Research Foundation, Unit Solar and Wind Energy, Jos Beurskens, Westerduinweg 3, PO Box 1, 1755 ZG Petten, The Netherlands

ECOFYS, Tony Schoen, Kanaalweg 16-G, 3526 KL Utrecht, The Netherlands

Folkecenter for Renewable Energy, P. Maegaard, Kammersgaardsveij 16, Sdr. Ydby 7760 Hurup Thy, Denmark

FZJ – Forschungszentrum Jülich, Gerd Eisenbeiß, Leo Brandt Strasse, 52425 Jülich, Germany

Fraunhofer ISE – Institute for Solar Energy Systems, Joachim Luther, Heidenhofstrasse 2, 79110 Freiburg, Germany

Genec Cadarache, Philippe Malbranche, Patrick Jourde, Bâtiment 351, CE Cadarache, 13108 St. Paul les Durance Cedex, France

HMI – Hahn-Meitner Institut, M. Lux-Steiner, Glienicker Strasse 100, 14109 Berlin, Germany

HUT – Helsinki University of Technology, Advanced Energy Systems/NEMO, Peter Lund, Otakaari 3 A, O2150 Espoo, Finland

IEE – Institut für Elektrische Energietechnik, Universität Kassel, Werner Kleinkauf, Siegfried Heier, Wilhelmshöher Allee 71/73, 34121 Kassel, Germany

IES-UPM – Instituto de Energia Solar, Universidad Politécnica de Madrid, Antonio Luque, Ciudad Universitaria, 28040 Madrid, Spain

IMEC – Interuniversity Microelectronics Center, Robert Mertens, Johan Nijs, Kapeldreef 75, 3001 Leuven, Belgium

ISET – Institut für Solare Energieversorgung e.V., Jürgen Schmid, Königstor 59, 34119 Kassel, Germany

ITER – Instituto Tecnológico Energias Renovables, Manuel Cendagorta Galarza Lopez, Poligono Industrial de Granadilla, 38611 Grandilla, Tenerife, Spain

IT Power Ltd., Bernard McNelis, The Manor House, Chineham Court, Lutyens Close, Chineham, Hampshire, RG24 8AG, United Kingdom

JRC-EI – Joint Research Centre, Environment Institute, Heinz Ossenbrink, Via Enrico Fermi 1, 21020 Ispra (VA), Italy

NMRC – National Microelectronics Research Centre, Gabriel M. Crean, University College, Lee Maltings V.C.C., Prospect Row, Cork, Ireland

NPAC – Newcastle Photovoltaics Application Centre, Nicola Pearsall, Ellison Place, Newcastle upon Tyne, NE1 8ST, United Kingdom

NTUA –RENES – National Technical University of Athens, Renewable Energy Sources Unit, Arthouros Zervos, PO Box 64011, 15701 Zografou, Athens, Greece

RAL – Rutherford Appleton Laboratory, Energy Research Unit, Jim Halliday, Building R63, Chilton, Didcot, Oxon OX11 0QX, United Kingdom

RISØ, Department of Wind Energy and Atmospheric Physics, E.L. Petersen, PO Box 49, 4000 Roskilde, Denmark

UA – University of Athens, Department of Applied Physics, D.N. Assimakopoulos, M. Santamouris, Physics Building V, 157 84 Zografou, Greece

UCD – University College Dublin, Energy Research Group, Owen Lewis, Donald Fitzmaurice, School of Architecture, Richview, Clonskeagh, Dublin 14, Ireland

Universidad Nova de Lisboa, Faculdade de Ciencias e Tecnologia, L. Guimarães, Elvira Fortunato, Quinta da Torre, 2825 Monte da Caparica, Portugal

Universidad Politecnica de Cataluna, GDS, ETSI de Telecomunicacion, L. Castañer, Modulo C4 Campus Nord, Calle Jordi Girona 1, 08034 Barcelona, Spain

Uppsala University, Claes-Göran Granqvist, School of Engineering, Dept. of Material Science, Solic Stae Physics, PO Box 534, 751 21 Uppsala, Sweden

VTT Energy, Kai Sipilä, PO Box 1601, 02044 VTT Espoo, Finland

WIP – Renewable Energy and Environmental Technologies, Peter Helm, Sylvensteinstrasse 2, 81369 München, Germany

ZSW – Zentrum für Sonnenenergie- und Wasserstoffforschung, Friedel Oster, Hessbrühlstrase 21C, 70565 Stuttgart, Germany

APPENDIX B.

PHYSICAL UNITS AND CONVERSION FACTORS

W	Watt
Wh	Watt hour
W_p	Watt Peak (see photovoltaics)
W_e	Watt of electrical power
W_{th}	Watt of thermal power
toe	Tonne of oil equivalent
J	Joule

1 TWh = 0.086 Mtoe (energy equivalent)
1 TWh = 0.222 Mtoe (corresponding to a Carnot efficiency of 38.46 % for the conversion of fuels into electricity)

PREFIXES

k	Kilo- (10^3)
M	Mega- (10^6)
G	Giga- (10^9)
T	Tera- (10^{12})
P	Peta- (10^{15})

INDEX

absorbed 101
active heating 80
adsorbed 101
advanced windows 81
AEAM (Association of European
    Automotive Manufacturers) 22
air-to-ground heat exchanges 81
amorphous-silicon 36, 47
    key issues and challenges 47
research and development goals 48
anaerobic digestion 9, 20
angle selective coating 105
availability 147, 154

bio-coal 4
bio-diesel 4, 19, 20
bioelectricity 4, 6, 7, 9, 14
bioethanol 4, 21
biofuels 4, 6, 7, 9, 14, 21
biogas 4
bioheat 4, 6, 7, 9, 14
biomass
    byproducts 4
    conversion 3, 5, 9, 12, 13, 14
    economic effects 6
    end-uses 3, 5
    energy 4, 5, 7, 8, 10, 12
    environmental effects 6
    in developing countries 198–199, 204
    list of policies 14
    pollutants 5
    potential 5
    production chains 9
    research 8, 9, 11, 12, 13
    resources 3, 5
    sources 5
    techno-economic aspects 6, 8, 9
    technological effects 6, 8
biomaterials: see bioproducts

bio-oils 4
bioproducts 4
bioresource 4, 12
blade pitch control 144, 146, 147, 162
boilers 15, 16
Brayton cycle 123, 136
bulb turbines 211

capacity credit / factor 153, 164
CdS cells 51
CdTe modules 32, 51
cell processing 43–44
    cooling of material 27
    Czochralski growth 27
    edge defined film growth (EFG) process 39
    environmental aspects 44
    ingot formation 27–28, 42
    slicing silicon 43
    wafer production 42
central receiver 115, 116, 123, 128, 232
CHP 16, 17
CIS modules 32
Clean Development Mechanism 143
cogeneration 14, 17
combustion of biomass 9, 15
concentrators and photovoltaics 41, 57–58
controllability 142
conversion technologies for biomass 6
copper-indium-diselenide cells 49
coriolis forces 208
cost reduction potential 31
cross-flow turbines 212
cut-in wind speed 154
cut-out speed 154
Czochralski growth 27

daylighting 81, 95, 104
dessicant cooling 108
direct solar steam generation 121, 122, 132, 135

direct-drive generator 162, 163
DISS 121

electrical characteristics of a solar cell 63–64
electricity storage for photovoltaics 36, 62
electrochromic glazing 97
electronic control in small hydro power 75
energy
    and load management 181
    autarky 83
    crops 5
    demand 195–197
    storage and renewables 181
    sustainability 83
environmental aspects of photovoltaic
    production 44
environmental effects of small hydro
    power 76
environmental externalities 184
ESHA (European Small Hydro Association) 65
ETBE 22
European RE integration strategy 187–193
European Wind Atlas 139, 140, 165, 167
European Wind Energy Association
    142, 144, 152
EWEA 142, 144, 152

feedstock 5, 8, 9, 12
fish protection in small hydro power 76
fixed tariff 151
Fix-Focus collector 225, 228
flat plate collector 223
fluidized bed systems 15
fuel cells 14, 83
funnel 142
furnaces 15
fuzzy logic 91

gas cleaning 9
gasification process 9, 16, 17, 18
gasochromic 105–106
GEF 118, 133
glare 95
Global Environment Facility 118, 133
green certificate 151
green funds 89
g-value 86, 97

harmonic distortion 142, 148, 163, 166
head enhancement in small hydro power 75
heat pumps 81, 83, 95

heat-pipe receiver 232
heliostat 123, 124, 126, 128, 132, 135, 232
high-head hydro 70
HVAC 93, 102
hybrid 148, 149
hybrid solar / fossil 115, 119, 132, 222
hybrid systems 180
hydro power 65

IAEA 152
IEA 142
III-V cells 52–53
induction generators in small hydro power 75
inflatable weirs 75
integrated performance tools 96
integrated performance view (IPV) 96
integrated RE systems 177
    classification 178–179
    economics 183–185
Integrated Solar Combined Cycle
    Systems 119, 120, 122, 123, 125, 126, 128,
    132, 135
integration in the urban environment 177
intermittent energy output 180
internal gains 80
International Atomic Energy Agency 152
International Energy Association 142
inverter 29
IPV 96
ISCCS 119, 120, 122, 123, 125, 126, 128, 132,
    135

joint implementation 143

La Rance tidal barrage 215
lean (building) 81, 82
lean climatization concept 94, 99–100, 104
legislative framework for renewable
    energies 192
LEO 93
levelized cost method 154
LIGHTSCAPE 95
ligno-cellulosic crops 5
liquefaction 19
load factor in small hydro power 69
Low Energy Office 93
low-head hydro 65, 70,
low-e coating 86
low-head hydro 211
LS-3 trough 121, 122

marine biomass fuel 208, 211, 216, 217, 220
marine currents 208, 211, 218, 220
market liberalisation 182, 186
match flow technique 91
metallic membrane collector 228
micro-hydro 65
mini-hydro 65
molten salt-in-tube receiver 123, 124, 127, 135
municipal solid waste 5

near-shore 152, 170
neural network 91

ocean energy
    economic aspects 217
    environmental aspects 217
    European projects 212, 214
    goals 218–220
    RD&D 218
oceanic wave climate 210
offshore 139, 141, 142, 144, 146, 151, 160, 161,
    163, 167, 168, 170
oil crops 5
opaque 103, 105
organic cells 55–56
osmotic energy 208, 211, 216
OTEC (ocean thermal energy conversion) 208,
    210, 217, 220
output forecasting 160
OWC (oscillating water column) 213, 214, 217

parabolic dish 115, 116, 128–131
parabolic trough 115, 116, 120, 128, 134, 222,
    223, 232
phase change materials 81, 108
PHOENICS 95
photocatalytic 231, 239
photochemical 231, 233, 239
photochromic 106
photovoltaic cell production 27
photovoltaic conversion 26
photovoltaic systems
    batteries 58
    charge regulators 58
    inverters 59
photovoltaics
    advanced research 54
    advantages of crystalline silicon
       technology 40
    advantages of PV electricity generation 30
cost reduction potential 31

electricity storage 36, 62
    future 37
    goals for research 34–36
    in developing countries 199–200, 204–205
    issues of crystalline silicon technology 39
    module 29, 56–57
    mono-crystalline 32
    multi-crystalline 32
    R&D roadmap 38–39
    silicon shortage 40
    third generation 27
phytomass 4
pinch point analysis 227
point absorbers 213, 217, 219
policy environment for biomass 6
biomass conversion technologies 14–22
    anaerobic digestion 9, 20
    cofiring 9
    combustion 9, 15
    esterification 19
    gasification 9, 16, 17, 19
    liquefaction 19
    pyrolysis 18–19
polycrystalline thin-films 49
polymer cells 55–56
power quality 148
predictability 142, 165, 166
preheaters 81
primary energy 83
private economy method 154
production chains of biomass 8
pyrolysis process 18–19

RADIANCE 95
radiant heating 236
Rankine cycle engine: see vapour cycle engine
reactive power 142, 148, 163, 166
refuse 5
residues 5
rural electrification 202–204

salinity gradient 208, 211, 216, 217, 220
SEGS 117, 118, 122, 133
self-cleansing 106
self-sufficiency 84
semi-transparent 103
sewage sludge 5
short rotation forestry 5
silicon 27
    electronic grade (EG) 27
    metallurgical grade (MG) 27

siphon turbines in small hydro power 74
small hydro in developing countries 201, 205
socio-economic issues of renewable
    energies 189
solar
    building 81
    cells 81
    collectors 95, 102
    combi-system 90, 108
    contracting 104
    control 97
    district heating 89
    fuel 236
    furnace 232
    gains 80, 82, 85, 103
    thermal power in developing countries
       201, 205
    transmittance 86, 97
    wall 103
'solar city' concepts 177
sorptive cooling 94, 101–102
stall control 144, 146, 147
starch crops 5
stretched membrane 124, 129
submersible turbo-generators in small hydro
       power 74
substrates of photovoltaics 43
sugar beet 5
sugar crops 5
switchable glazing 97

thermal conversion 26
Thermie programme 20
thermochemical 232, 235
thermochromic coating 105
thermotropic coating 105
third-party financing 89, 104
tidal barrages 210–211, 215–216, 217, 220
tidal currents 208, 211, 218, 220
total life-cycle balance 83
transparent insulation 87
turbines in hydro power
    impulse 68
    reaction 68
    sizing 68
    small-scale 65, 67
    variable speed 74

U-value 86, 97

vacuum tube 223
vapour cycle engine 210, 215
volumetric air receiver 124, 126, 135
volumetric receiver 232

wafer production 42
waste air heat pump 91
wastes 4
wave energy 208, 209, 213–215, 218, 220
WEC (wave energy converters) 213, 219, 220
Wells turbine 213
wind farm 147
wind in developing countries 202, 205
wood processing waste 5, 6